XII All-Russian Scientific and Practical Conference on Automation Systems in Education, Science and Production 2019

IOP Conference Series: Materials Science and Engineering
Volume 865

Novokuznetsk, Russia
28 - 30 November 2019

ISBN: 978-1-7138-2688-0
ISSN: 1757-8981

Printed from e-media with permission by:

Curran Associates, Inc.
57 Morehouse Lane
Red Hook, NY 12571

Some format issues inherent in the e-media version may also appear in this print version.

Institute of Physics
Dirac House, Temple Back
Bristol BS1 6BE UK

Phone: 44 1 17 929 7481
Fax: 44 1 17 920 0979

techtracking@iop.org

Additional copies of this publication are available from:

Curran Associates, Inc.
57 Morehouse Lane
Red Hook, NY 12571 USA
Phone: 845-758-0400
Fax: 845-758-2633
Email: curran@proceedings.com
Web: www.proceedings.com

TABLE OF CONTENTS

Preface

Since 1995 Siberian State Industrial University (Novokuznetsk) together with research and design institutes, educational organizations and industrial enterprises, socio-economic structures has been holding bi-annually the All-Russian scientific-practical conference (with international participation) "Automation Systems in Education, Science and Production". The conference is covers a wide range of issues on the creation and improvement of automation systems by industrial and socio-economic facilities.

The program and Organizing Committee of the conference includes well-known domestic scientists and developers of automation systems, as well as representatives of the CIS countries. The conference is supported financially (on competitive terms) by the Russian Foundation for Basic Research and large industrial associations, such as LLC "United Company Sibshakhtostroy" (Novokuznetsk), CJSC "Stroyservis" (Kemerovo).

In 2019, the conference was held on November 28-30. The conference organizers received 92 papers from specialists from Moscow, Ekaterinburg, Omsk, Novosibirsk, Kemerovo, Krasnoyarsk, Prokopyevsk, Novokuznetsk, Magnitogorsk, Novotroitsk, Lipetsk, Kaliningrad, Samara, Barnaul, Bishkek, Almaty, Irkutsk, Angarsk, Abakan, St. Petersburg, Rostov-on-Don, representing academic, research, educational and design institutes, administrative and socio-economic structures, enterprises of various profiles.

The conference was organized in sections: plenary session, current problems of the theory and practice of automation, automation and informatization systems for industrial purposes, educational purposes, research purposes. A round table was organized on the theme: "Cross-cutting technologies. Solutions for predictive equipment maintenance". In the conference sections students, graduate students, young scientists as well as specialists from industrial enterprises of Kuzbass took an active part.

The Organizing Committee thanks LLC "United Company Sibshakhtostroy" (Novokuznetsk), LLC "ATESKO Sibir" (Novosibirsk), LLC "Scientific Research Center for Control Systems" (Novokuznetsk) for organizational and financial support, as well as all Conference participants for their contribution to the development of the theory and practice of control automation.

The Chair of the Conference Organizing Committee
Doctor of Technical Science, Professor,
Rector of SibSIU E.V. Protopopov

IOP Conf. Series: Materials Science and Engineering **865** (2020) 011002 doi:10.1088/1757-899X/865/1/011002

Peer review statement

All papers published in this volume of *IOP Conference Series: Materials Science and Engineering* have been peer reviewed through processes administered by the proceedings Editors. Reviews were conducted by expert referees to the professional and scientific standards expected of a proceedings journal published by IOP Publishing.

IOP Conf. Series: Materials Science and Engineering **865** (2020) 012001 doi:10.1088/1757-899X/865/1/012001

Control systems for complex technological facilities with the accumulation of experience in the development of control actions

S M Kulakov, M V Lyakhovets, R S Koynov and E N Taraborina

Siberian State Industrial University, 42 Kirova Street, Novokuznetsk, 654007, Russia

E-mail: kulakov-ais@mail.ru

Annotation. The complex task of man-machine control of weakly formalized technological units and complexes, which are characterized by a wide variety of conditions, multidimensionality, non-stationarity, uncertainty, multivariate products, is considered. The insufficient effectiveness of the model approach to creating control systems for such objects is shown. Alternative approaches based on the concept of best practices are considered. The well-known best practice procedures using typical representative situations and the model cycle method are presented. A new (precedent) method for the selection and implementation of control actions with the participation of an operator-technologist is proposed. A precedent cycle for selecting controls and a functional diagram of the control system have been developed. An information model of the precedent is formed on the example of steelmaking control in an oxygen converter. The advantages of the precedent method compared with the method of representative situations are determined.

1. Introduction

The problem of automated control of weakly formalizable technological objects (WFTOs), which include many technological units and human-technical complexes of various industries, today cannot be recognized as effectively solved on the basis of the traditional (model) approach. The essence of the latter consists in the identification of the control object, that is, in the construction of a mathematical model of control channels and influence channels of the controlled disturbing influences of the object, and in the case of human-technical complexes, in the construction of regulatory models and the synthesis on their basis of algorithms (mechanisms) for the generation and implementation of control actions.

The lack of effectiveness of the model approach to the creation of automated control systems of WFTO is due to: the intensity, complexity, multiplicity and unsteadiness of internal processes; incompleteness and errors of control of numerous parameters; multi-mode operation, multivariance of products; the variability of the characteristics of the object during operation. The listed properties and features of WFTO, in combination with the lack of effectiveness in managing them, prompt us to look for other approaches to making control decisions. For example, it is useful to turn to the modification and algorithmization of the concept of "best practices" that is widely known and used in organizing people's activities, as applied to the management practices of WFTO [1, 2]. In the field of control of technological units and complexes of ferrous metallurgy units in this direction, two research directions can be distinguished: [3, 4, 5], [6,7]. This report briefly analyzes the traditional (model) approach to the construction of automatic control systems [8, 9], [10], [11, 12], gives a description of the two above full-scale development models that specify the best practices for managing WFTO in metallurgy and, more the author's approach to the algorithmization of accumulation of experience in the control of WFTO based on the precedent method is presented in detail [13, 14, 15].

2. The traditional model approach to the construction of automatic process control systems

A distinctive feature of the classical method of synthesis of a feedback control algorithm is the use of a mathematical model of the object's channels, as well as the influence model of the reduced uncontrolled disturbance. In a simple case, the structure of the algorithm is selected from among the typical ones (P, I, PI, PID, etc.), and its settings are determined on the basis of empirical formulas or by solving the optimization problem by the criterion of the minimum variance of control errors, Figure 1, [8, 9].

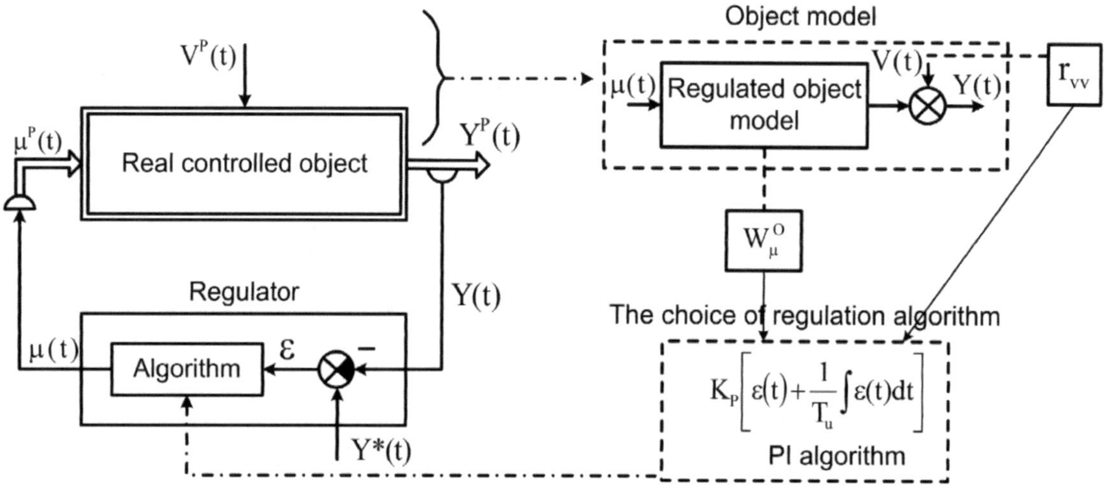

Figure 1. The classical scheme of selection of the regulation algorithm.

Designations: $Y^P(t), \mu^P(t), V^P(t)$ - real output action, control action, uncontrolled disturbing influences of the object at the moment t; $Y(t), Y^*(t)$, $\mu(t)$ - measured and set output, as well as the control action of the object at the moment t; $V(t)$ - equivalent disturbance reduced to the output of the object; r_{vv}, W^O_μ - autocorrelation function and object operator; K_P, T_u - tuning parameters of the PI algorithm.

The original development of SibSIU Professor V.P. Avdeev and his colleagues is PPR - a predictably-predictable regulator, [10], the structure of the algorithm of which explicitly includes a model of a control channel, Figure 2.

Designations: $U^\partial, Y^\partial, \omega^\partial, \varphi^\partial_O, \varphi^\partial_\tau$ - actual control, output, disturbing influences, as well as the operator of the control channel and the operator of delay; Y, U^B_τ, U^Π - output variable, restored ideal control action, predicted control action; $\varphi^M_O, \varphi^M_\tau$ - model operators of the control and delay channel; $[\varphi^M_O]^{-1}$ - reverse operator of the control channel model; f^Π - prediction operator; , - the symbol of the actuator and the measuring information sensor.

The main operations of PPR are: estimation of the deviation of the controlled variable from the task, retrospective restoration of the ideal control action (using the inverse model of the control channel), correction of the control taking into account the delay operator, extrapolation of the time series of ideal control actions, implementation of the extrapolated control action.

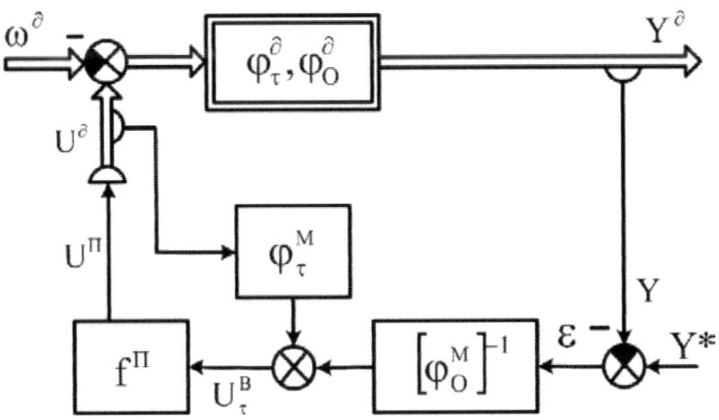

Figure 2. The structure of a simplified PPR (without disturbance compensator and corrective circuit).

The so-called Advanced Process Control (APC) systems are considered more efficient in relation to typical systems and more complex, that is, systems for the advanced control of multidimensional technological objects, the main component of which is the predictive object model and predictive control algorithms (Model Predictive Control - MPC) , [11,12]. They are widely used in oil refining, chemical, pulp and paper and other enterprises in the world. A control system diagram with a predictive model is shown in Figure 3. With respect to the PID controllers, MPCs give an effect in these enterprises in the range of 0.5-5.0 million dollars per installation per year, [12]. The control action selection algorithm uses numerical optimization to find controls U^H on the control horizon using forecasts Y^H for the dynamic memory interval of the object.

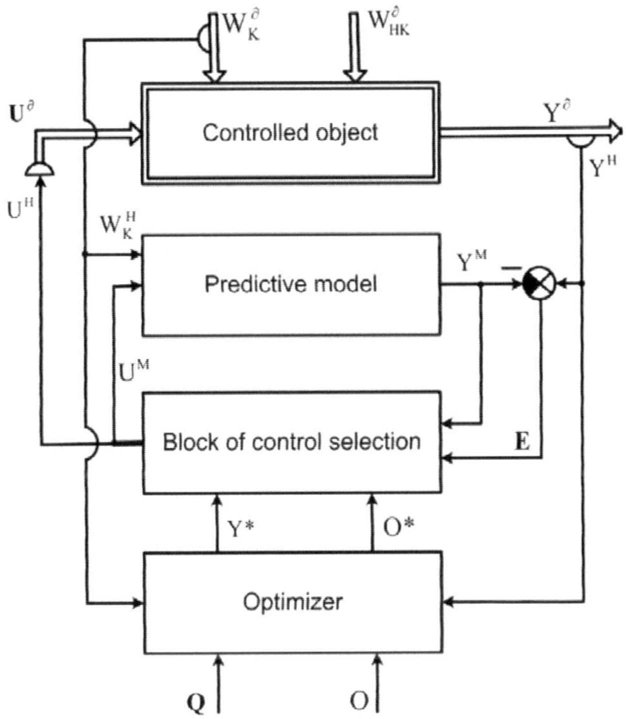

Figure 3. The structure of the control system with a predictive model.

Designations: $U^\partial, Y^\partial, W_K^\partial, W_{HK}^\partial$ - actual control, output, as well as controlled and uncontrolled disturbing influences (disturbances) of the control object; U^H, Y^H, W_K^H - full-scale control signals, output signals of the object and signals of controlled disturbances; U^M, Y^M - model control signals and output signals of the object; Y^*, O^* - set values of the output effects of the object and restrictions on the controls; Q, O - objective functions and limitations of the optimizer.

The considered effective structures of automatic control systems for technological objects are characterized by implicit [8, 9] or explicit [10, 11, 12] using a functional or essential model of the control object, which is adequate for a long time or can be updated in a timely manner (with a significant change in the properties of the object) with using the identification subsystem integrated into the control system. Such systems are ineffective for poorly (poorly) formalizable objects, for example, coke oven complexes, blast furnaces,

IOP Conf. Series: Materials Science and Engineering **865** (2020) 012001 doi:10.1088/1757-899X/865/1/012001

oxygen steelmaking converters and arc steelmaking furnaces. It is advisable for them to create control systems based not on the model, but on the full-scale and (or) full-scale approaches, [16].

3. Characterization of known methods and control systems WFTO with control experience

3.1. Set-based forecasting and control

In [3, 4, 5], it was proposed to accumulate and use the experience of forecasting and decision making in control systems in the form of a set of typical representative situations (TRS). At the same time, TRS is understood as an interconnected set: structure, information display of an object, external and internal conditions of its functioning, parameters of control channels and controlled external influences, implementations of the given disturbing influences, control performance criteria.

The final set of TRS is considered as a full-scale model based on which control decisions and (or) forecasts are made. For example, more than 50 TRS are used in the predictive control system of the coke-chemical production of a metallurgical plant, [5].

The procedure for selecting controls based on TRS includes the following main actions: collecting data on the current situation in the control system; recognition of the class to which the current situation relates; adoption of an effective control action (or forecast) recommended for the selected class of situations; correction of the recommended impact (forecast), taking into account the differences in the characteristics of the current situation from the characteristics of a typical representative of the class, the implementation of the formed control action. The introduction of the TRS method is presented in detail in [5], which is a clear example of the effective control of WFTO.

3.2. Control of the unit of cyclic action on exemplary cycles

In [6, 7], a concrete example of the application of the control method for model cycles, namely, control of steel melting in an oxygen converter for model melts, is considered. All swimming trunks became divided into classes depending on their input (initial) and output (final) variables. The classification of steel melts involves the use of the following elements: the content of silicon, manganese, phosphorus, sulfur in molten iron; temperature and mass of molten iron; the content of carbon, phosphorus, sulfur in steel; steel temperature; basicity of slag, minute consumption of blast; position of the purge lance. The range of values each of these variables is divided into several (up to 5) subranges. Such a partition corresponds to no more than 65 classes. If the input and predetermined output values of the variables of the forthcoming melting coincide with one of the previously produced exemplary (effective) melts, information on the parameters of which is stored in the database, then the values of the control actions for the forthcoming melting are taken to be the same as in the found model melting.

This method is basically similar to the method of making control decisions on typical representative situations, TRS. But it does not imply the operational adjustment of the controls using the recalculation model (the model of the object in “small”), and is an example of a full-scale approach to the control of WFTO, which includes the process of steelmaking in the converter.

4. A case-based method for the accumulation and use of decision-making experience in the man-machine control system of WFTO

The essence of the case-based decision-making method is widely known and clearly displayed by the so-called CBR-cycle of case-based decision-making, [13, 14]. At the same time, the algorithmic foundations of a precedent approach to making control decisions in automated (man-machine) process control systems are still at the initial stage of their development [15, 16, 17].

First of all, it is necessary to modify the CBR decision-making cycle, considering it as a cycle of generating control actions based on information about the situation in the control system (about external influences, parameters of the state of the unit, output actions, past and current controls, control goals and limitations). Figure 4 shows such a modified cycle of generating control actions on the basis of precedents (the case-control cycle - CPU).

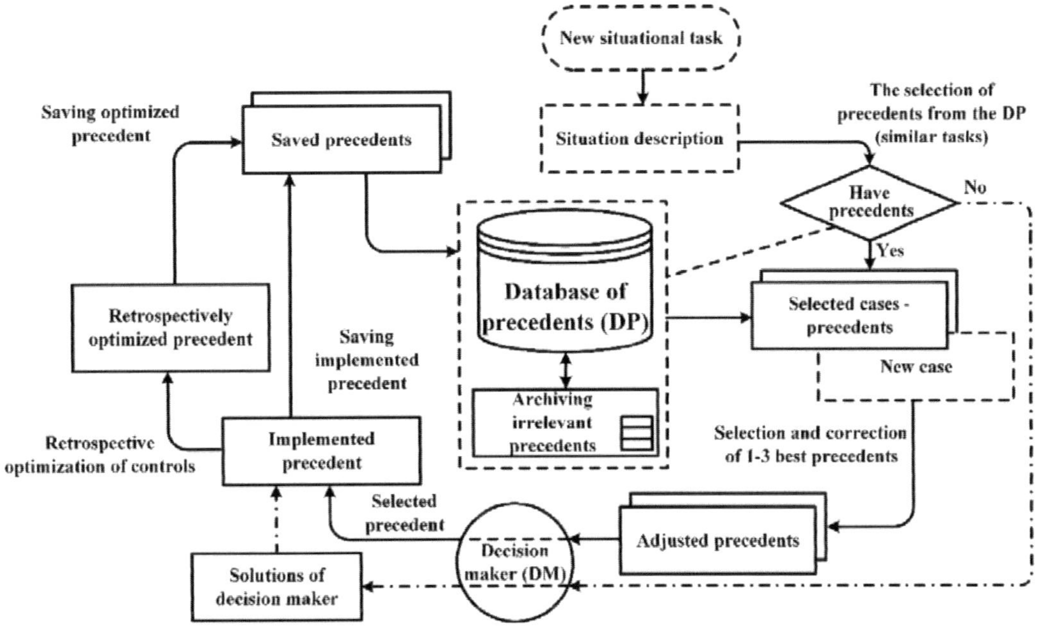

Figure 4. Modified cycle for the development and implementation of control decisions in the control system of WFTO.

Actual tasks related to the modified CBR cycle (CPU) are: the development of a functional diagram of the control system of WFTO; building a precedent information model, choosing the best (optimal) precedent from the set of relevant, at the moment, precedents and others. For example, the development of a control program for the forthcoming steelmaking in an oxygen converter, including the number and times of delivering charge materials, graphs of the flow rate of the blast and the position of the purge form, mass and rhythm of the supply of slag-forming materials, melt temperature, developed: general scheme of the control system, figure 5; information model structure of precedent, figure 6.

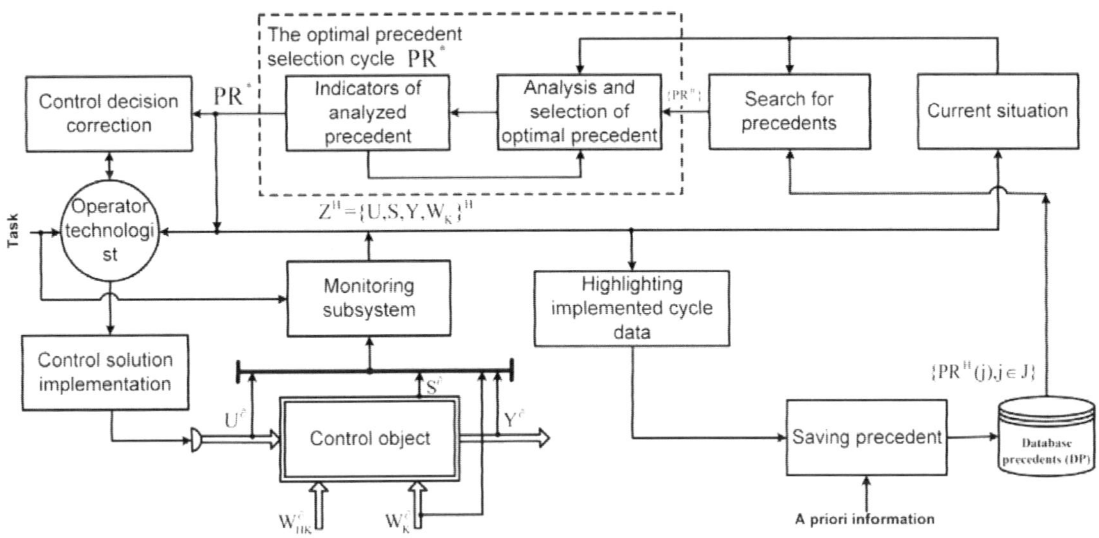

Figure 5. Scheme of the control process WFTO cyclic action.

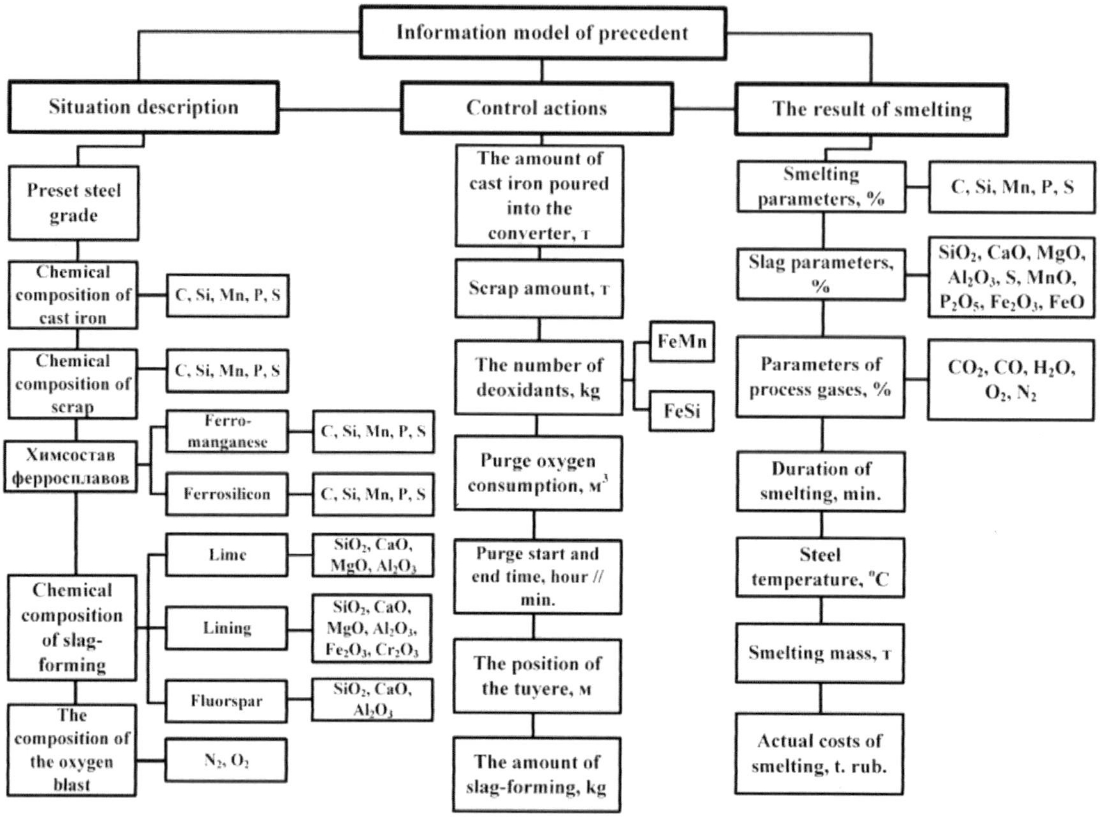

Figure 6. The structure of the precedent information model (for the synthesis of the control program for the upcoming steel melting in the converter).

5. Conclusion

An important issue with respect to decision-making by the use-case method is the identification of the similarities and differences of this method with respect to the TRS method and its special case – the model of model cycles. First of all, it should be noted that all three methods under consideration (TRS method, model cycle method, precedent method) belong to the same class of control decision-making methods. Namely, to the class of methods characterized by the accumulation and subsequent use of data on earlier implemented effective (as well as ineffective) controls, briefly to the class of methods with accumulated control experience (ACE methods).

The main differences between the TRS method and the TRS method are:

1) Different basic concepts: type representative (TRS), precedent.

A type representative is a characteristic, regularly occurring, typical object, process, situation, implementation of a measuring signal, a decision-making task.

A precedent is any specific decision-making case, process, event, situation that has taken place (implemented) in the past, similar to the upcoming decision-making case, process, event, event.

2) Different concepts of making control decisions (on the basis of TRS and on the basis of precedents).

To develop a control decision on the basis of TRS, it is necessary to form a set of typical situations and corresponding effective solutions, from which a decision is made for the current situation in the control system, after assigning it to a TRS.

In order to develop a decision based on precedents, it is necessary to create and update online many specific cases of decision-making (base of precedents), from among which a choice is made (and correction, if necessary) of the forthcoming decision that is as close as possible, by initial conditions, to the precedent optimal at the moment.

3) The power of many TRS in control systems of complex objects is always much (by orders of magnitude) less than the power of many precedents. Accordingly, the TRS method does not meet the principle of necessary diversity of U.R. Ashby in relation to WFTO, which are characterized by a very high variety.

4) Decision-making methodology based on precedents use new (with respect to decision-making based on TRS) tools, such as:

- domain ontology;
- theory of structural mapping;
- methods for constructing an information model of precedents;
- fuzzy precedent classification rules;
- methods for selecting optimal precedents;
- methods for assessing the relevance and archiving of actual precedents in real time.

References

[1] Bezginova Ju A 2018 *Otkrytoe Obrazovanie* **22(6)** 27–38
[2] Zimin V V et al 2013 *Fundamentals of Life Cycle Management Services for Informatics and Automation Systems (ITIL Best Practices): Tutorial* (Kemerovo: Kuzbassvuzizdat) p 500
[3] Avdeev V P et al 1984 *Types of Representative Offices in Research and Management Tasks* (Kemerovo: KemSU) p 91
[4] Emel'janov S V et al 2008 *Theory and Practice of Forecasting in Control Systems* (Kemerovo, Moscow: Rossijskie Universitety: Kuzbassvuzizdat – ASTSh) p 487
[5] Myshljaev L P 2006 *Industrial Automation Systems* vol 2 (Nauka) p 483
[6] Bogushevskij V S et al 2006 *Stal'* **1** 18—21
[7] Bogushevskij V S et al 2007 *Metallurgicheskaja i Gornorudnaja Promyshlennost* **4** 232-235
[8] Rotach V Ja 1973 The *Calculation of the Dynamics of Industrial Automatic Control Systems* (Moscow: Energiya) p 440
[9] Rotach V Ja 2008 *Theory of Automatic Control* (Moskva: MJeI) pp 396-400
[10] Avdeev V P et al 1989 *Restorative and Predictive Control Systems: Textbook* (Kemerovo: KemSU) p 91
[11] Carlos E Garcia et al 1989 *A Survey. Automatica* **25(3)** 335–348
[12] Dozorcev V M and Kneller D V 2005 *Datchiki i Sistemy* **10** 56-62
[13] Aamodt A and Plaza E *Case-Based Reasoning: Foundational Issues, Methodological Variations, and System Approaches. AI Communications. IOS Press* vol 7: 1 pp 39-59
[14] Varshavskij P R et al 2013 *Information Models and Analyses* vol 2 No 4 385-392
[15] Karpov L E et al 2007 *Trudy Instituta Sistemnogo Programmirovanija (RAN)* vol 13 part 2 pp 37-58
[16] Kulakov S M et al 2011 *Proc. of the VIII All-Russian Sci. and Pract. Conf. in Automation Systems in Education, Science and Industry* (Novokuzneck: SibSIU) pp 26-34
[17] Kulakov S M et al 2017 *Proc. of the XI All-Russian Sci. and Pract. Conf. in Automation Systems in Education, Science and Industry* (Novokuzneck: SibSIU) pp 11-19
[18] Kulakov S M et al 2019 *Trudy Vserossijskogo Soveshhanija po Problemam Upravlenija. VSPU-2019* http://vspu2019.ipu.ru

IOP Conf. Series: Materials Science and Engineering **865** (2020) 012002 doi:10.1088/1757-899X/865/1/012002

Modern concept of technological processes modal control in a state space combined with wavelet monitoring of current modes

D B Fedosenkov[1], A A Simikova[2] and B A Fedosenkov[2]

[1]Siberian Generation Company, 53 Dubininskaya Str., build. 5, Moscow, 115054, Russia
[2]T.F. Gorbachev Kuzbass State Technical University, 28 Vesenniaya Str., Kemerovo, 650000, Russia

E-mail : rafwaveletsve@mail.ru

Abstract. The article deals with the aspects of modal control in spaces of system states and wavelet time-frequency distributions. On the example of a mixture-producing aggregate mathematical models are formed to represent technological feeding material flow rate signals, and the relevant vector-matrix model based on the state space method is developed for the feeding devices unit (FDU). It is revealed that the signals generated by the feeders, are non-stationary in frequency and amplitude. They cannot be adequately interpreted by standard analysis techniques, e.g. such as the Fourier transform. To identify current operating modes of the feeding unit, it is proposed to use the apparatus of wavelet transforms, which allows you to identify local features of the flow signals in a combined time-frequency medium with essentially larger information richness and semantic clarity. For the aim of controlling current modes in FDU a closed automatic modal control system (CACS) with a full state vector feedback has been developed. In the modal control algorithm the feedback matrix is permanently recalculated in real time, which forms the specified localization of the poles constellation. Specified non-stationary feeding modes are demonstrated to represent the essence of modal controlling in the FDU.

1. Introduction

In technological schemes of manufacturing certain products for various purposes related to the production of multicomponent mixtures, the greatest efficiency of the feeding and mixing processes in the preparation of combined products is achieved in continuous-type mixture-producing aggregates. Moreover, regardless of the type of flow sensors, the signals generated by the feeders, are non-stationary in frequency and amplitude. Such signals cannot be adequately interpreted by standard analysis techniques, e.g. such as the Fourier transform.

As a solution to the problem of identifying the current operating modes of the mixing unit, it is proposed to use the apparatus of wavelet transforms, which allows you to identify local features of the flow signals both in time and frequency. As a result of the consideration of modeling methods for the processes of continuous preparation of mixtures, the expediency of studying their dynamics using the state space method, time-frequency wavelet analysis, and also methods to describe the feeding devices unit (FDU) as a controlled dynamic system, is revealed.

2. Modeling the feeding devices unit in a sate space

In order to stabilize material flows at a nominal level in the framework of the pre-mixing stage (it's necessary to obtain high quality mixtures, that is, mixtures with a low degree of heterogeneity), a modal control method for the feeding process is implemented in the work, which implies the creation of conditions for maintaining the required modes due to the directed specific action on the "poles constellation" of a closed feeding system with vector feedback by acting on the executive mechanisms of the feeders. In this regard, the problem of the so-called reverse (recovering) transient processes resulting from the forced structural and parametric non-stationary state of the system in case of sporadic uncontrolled changes in input – with respect to a feeding device (FD) – disturbances is solved.

When modeling the FDU in the state space, in order to form a vector-matrix model (VMM) based on the operator functions of the links, a system of differential equations written in Cauchy form and the output equation are compiled. In this case, derivatives of the first and higher orders are written in the form of phase state variables, and a transport delay is approximated by a Padet polynomial-power fraction or by a volume delay in the chain of ten non-periodic links. Presenting the set of variables in the form of vectors, and the collection of parameters in the form of corresponding matrices, we obtain a vector-matrix model of a feeder unit in the state space.

The final model of the feeding flow signals for the spiral and screw feeders of the third order in terms of the state space as a vector-matrix model has the form [1]:

$$\begin{cases} \dot{x}_1(t) = x_2(t) \\ \dot{x}_2(t) = -\omega_{d_1}{}^2 x_1(t) + X_{dm_1}\omega_{d_1}u(t) \\ \dot{x}_3(t) = X_{d0_1}u(t) \\ u_1(t) = x_1(t) + x_3(t) \end{cases} \qquad \begin{cases} \dot{x}_4(t) = x_5(t) \\ \dot{x}_5(t) = -\omega_{d_2}{}^2 x_4(t) + X_{dm_2}\omega_{d_2}u(t) \\ \dot{x}_6(t) = X_{dm_2}u(t) \\ u_2(t) = x_4(t) + x_6(t) \end{cases}$$

The feeding signal of a discrete feeder is described by a system of differential equations which in the state space will take the form:

$$\begin{cases} \dot{x}_7(t) = \dfrac{A_0}{2}u(t) \\ \left.\begin{aligned} &\dot{x}_{2l}(t) = x_{2l+1}(t) \\ &\dot{x}_{2l+1}(t) = -((q-1)\omega_b)^2 \cdot x_{2l}(t) + u(t) \end{aligned}\right\} \quad q = \overline{2,11},\ l = \dfrac{(1-(-1)^q)}{2} + 2q \\ u_b(t) = x_7(t) + \sum x_{2i}(t),\ i = \overline{4,12} \end{cases}$$

Here l – a normalized state variable index; q is the redefined full harmonic number.

3. Representation of scalar material flow rate signals in a wavelet medium

The requirement to obtain high quality mixtures (with a low coefficient of heterogeneity) implies a uniform (with small fluctuations) supply of the initial ingredients to the mixing apparatus from the feeding devices unit. Therefore, already at the feeding stage, it is necessary to continuously monitor and, accordingly, adjust current operating modes of the feeding devices. Therefore, one-dimensional (scalar) signals of feeding material flows are recorded by means of strain gauge and /or piezoelectric transducers. The obtained primary waveforms of flow rate signals are distorted with noise and non-stationary in frequency, and consequently, at any time they have randomly changing instantaneous dynamic spectra. It is not possible to display such spectra on a visual level by existing traditional methods.

In this regard, a new approach is implemented in the article, which is based on mathematical models and algorithms for approximating, identifying and correcting controlled variables in a certain space. The latter is created on the platform of Gabor wavelet functions [2] and time-frequency wavelet thesauruses formed on their basis. Moreover, the used information about the state of the object is displayed in the wavelet medium in the format of two-dimensional / three-dimensional Cohen's class quadratic distributions [3]. The Wigner and Choi-Williams distributions were accordingly used as the working and training ones.

The essence of the controlling and monitoring procedure is as follows. When the spectrum structure of the temporal signal vector changes, the time-frequency localization of the corresponding elements (atoms) on the Wigner map [2, 3] also changes, which should be visualized and registered by the monitoring complex, after which (if necessary) a corrective action is put to the feeding electric drive.

Considering that in the controlled plant individual feeders sporadically (randomly, from time to time) change their output signals (due to variation in input actions arising owing to spontaneous changes in the physical and mechanical characteristics of the ingredients entering the controlled plant), it seems appropriate to observe the general output signal of the controlled plant flow, the wavelet-card of which captures all the signals of individual feeding devices. At the same time, on the operator monitor "nominal" frames (specific window frames – see figure 1) are formed, onto which time-frequency atoms of a wavelet distribution (for example, Wigner distribution) that correspond to the nominal feeding mode, are projected. For continuous-type feeding devices (CTFDs) of spiral and screw type, such distributions with permanent frequencies are thin in frequency (along the ordinate axis) and elongated in time (along the abscissa axis) time-frequency atoms (TFA) on a Wigner map. Therefore, under real operating conditions of feeders, on the wavelet map (W-map), a deviation of a certain TFA (for a CTFD) or a group of atoms (for a discrete-type feeding device – DTFD) sporadically takes place beyond the boundaries of the frame / frames. In this case, the computer control system records the sequence of violation of the rated modes of certain FDs. And as to similar modes, the processing time of the corresponding FD W-map should be less than the time interval between the occurrence of two adjacent-in-time deviations of the TFA outside the frames of two any different FDs or the same feeder.

When fixing the unacceptable deviation of the TFA beyond the borders of the frames, the modal control technology begins to operate.

Feeding processes are controlled by the computer system in a wavelet medium based on a modal control algorithm. The purpose of the algorithm is to stabilize the feeding process at the pre-mixing stage at a nominal level with fluctuations in the output flows owing to sporadic stepwise or exponential changes in the input actions on the feeders.

The essence of designing a modal control system is the synthesis of a closed automatic control system (CACS) with a full state vector feedback by certain placing the poles of this system, followed by determining the feedback matrix (the modal controller parameters), which forms the specified localization of the poles constellation. The synthesis by placing the poles is based on the use of the system model in the state space.

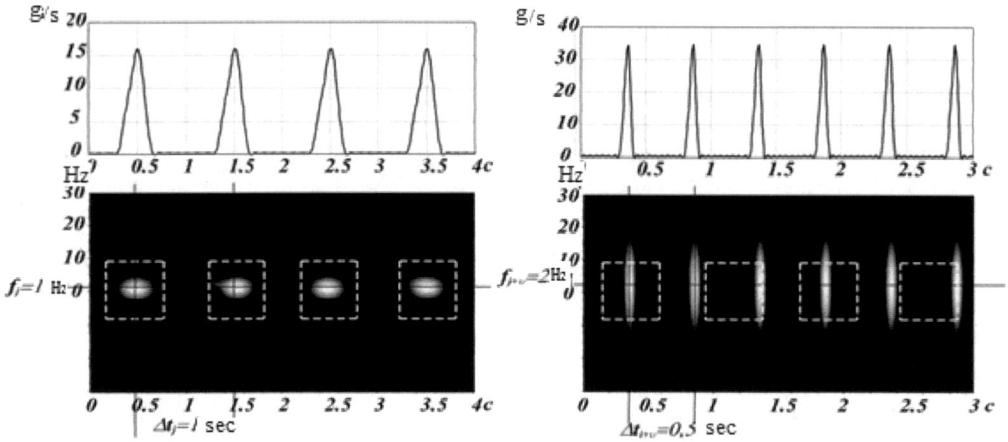

(a) (b)

Figure 1. Waveforms and their time-frequency maps corresponding to a) nominal and b) current (indignant) operating mode for a discrete-type feeder.

In real manufacturing feeding systems, the hopper containers of each FD are filled with the materials of the initial ingredients, which have different physical and mechanical properties in each tank (for example, moisture characteristics and / or uniformity degree of particles in composition), therefore during feeding there occurs sporadic variation of the effect of the fed ingredient material on the input part (i.e. material intake part) of the relevant feeder active organ. As a result of this, the frequency of the feeding signal (associated with the rotational speed of the FD executive mechanism – FDEM) changes in transient modes, which ultimately leads the feeding process to a new steady-state frequency mode that does not correspond to the nominal one required at the pre-mixing stage.

Upon receipt of a more uniform and dry (less humidified) substance, the FDEM accelerates due to load shedding on the feeder working body, while the frequency of the material flow signal increases until it stabilizes at a new steady level with a new, non-nominal, performance. But with an increase in the load on the working body of the feeding device the picture is opposite. In both cases, it is required to act on the CACS with modal control in order to stabilize the material flow at a nominal level. But since the action at the intake part and the reaction at the output of the FD have already stabilized at a new level, it is necessary to bring influence to bear upon the internal structure of the CACS in such a way as to change the structural and parametric properties of the system. Thus, it is required to create the effect of a forced structural and parametric non-stationary state by changing the localization of the CACS "poles constellation".

Consequently, during FDEM-speeding up (that is, when the signal frequency increases due to a decrease in the load (L-dump) on the FD working body), it is necessary – in order to stabilize the flow rate at the nominal level – to reduce accordingly the value of the imaginary conjugate poles pair (for CTFD). As for the reverse transient feeding process in DTFD, the imaginary poles should be reduced only for the most powerful sub-harmonics of the first harmonic in the Fourier model. In this case, the model has the form as follows: $B_1 \mathrm{Sin}\ \omega_1 t + A_1 \mathrm{Cos}\ \omega_1 t$. High harmonics are rejected using low-pass filtering or a high-frequency rejecting filter.

4. Transient processes under L-rise / L-dump

When implementing an upward reverse transient process after L-rise (URTP) is carried out in the frames of modal control, restoration of the nominal feeding mode is performed according to the two-exponential function of the form

$$x(t)_{b/ei} = x(t)_{0b/ei} + x(t)_{mb/ei} \sin[2\pi f_{var}(t - \tau_b) - \varphi_0],$$

where $f_{var} = f_b[1(t - \tau_b)] + (f_e - f_b)\ ([1(t - \tau_e)] - \exp[-(t - \tau_b) / T_3]\ T_3 / (T_3 - T_4) +$
$+ \exp[-(t - \tau_b) / T_4\ ]\ T_4 / (T_3 - T_4)$ – the instantaneous frequency of the chirp-signal (a signal with a time-dependent frequency) of the upward reverse transient process;

$x(t)_{mb/ei} = x_{mbi} +(1 / T)\ t,\ \ \tau_b \leq t \leq \tau_e$ – changing in the amplitude of the variable component of the URTP;

$x(t)_{0b/ei} = x_{0bi}\ [1(t - \tau_b)] + (x_{0ei} - x_{0bi})\ ([1(t - \tau_e)] - \exp[-(t - \tau_b) / T_3]\ T_3 / (T_3 - T_4) +$
$\exp[-(t - \tau_b) / T_4\ ]\ T_4 / (T_3 - T_4))$ – the non-periodic component of the upward reverse TP.

Thus, a complete transient graph function for two cycles of load variation (L-rise and L-dump), including a downward forward (DFTP) / upward reverse (URTP) transient processes, as well as an upward forward (UFTP) and a downward reverse (DRTP) transient processes, has the form (figure 2); here the ordinate axis is the flow rate one x (t). Figure 3 shows separately the graph functions of the upward and downward reverse TPs.

The procedure to implement the modal control algorithm for the feeding process including the technology of forced localization and relocalization of open ACS poles under the conditions of non-stationary processes in the "FDEM-FD" system, is considered on the example of a continuous-type feeder in a disturbed mode during load shedding (during a transient process like DRTP).

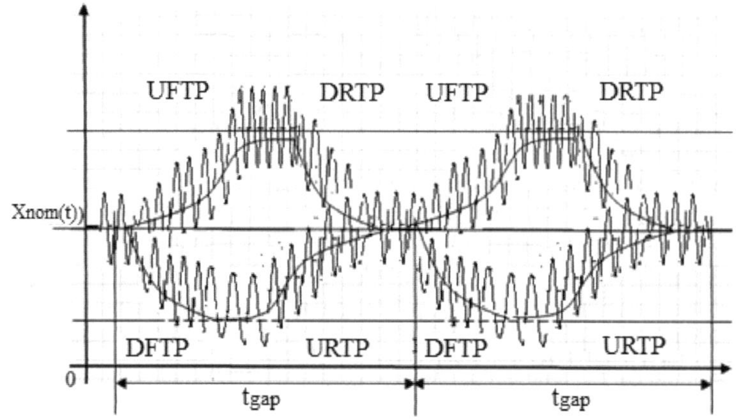

Figure 2. Cycles of modal controlling the feeding process under L-rise and L-dump.

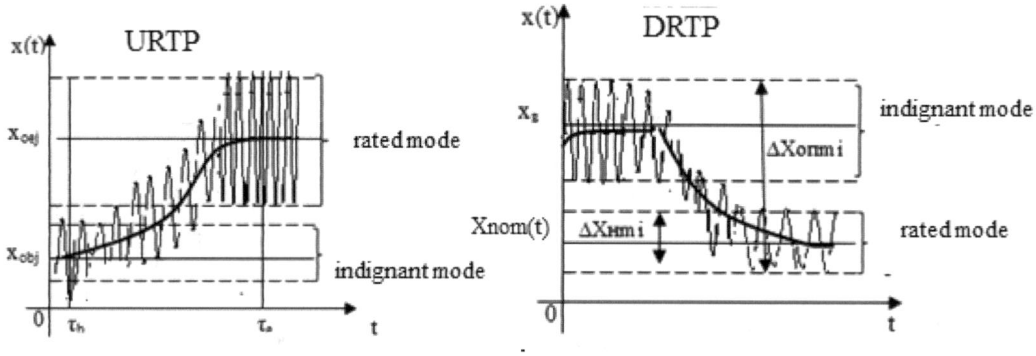

Figure 3. Waveforms of modal controlling the feeding process after the L-rise (on the left – URTP) and after the L-dump (on the right – DRTP).

5. The procedure of calculating the modal controller parameters

The calculation of the modal controller parameters, that is the determination of the feedback matrix K over the full vector of state variables, is reduced to its calculation by the Ackermann formula [4]:

$$K = \begin{bmatrix} 0 & 0 & 0 & 1 \end{bmatrix} \cdot Q_c^{-1} \cdot \alpha(A),$$

where Qc^{-1} – inverse controllability matrix: $\alpha\ (A)$ is the matrix polynomial formed according to the Cayley–Hamilton theorem [5, 6].

This matrix is continuously recalculated in real time with a time step of 10 ms during the transient process to ensure its time-frequency (chirp-) nature.

Using similar procedures, the computer control system calculates the modal controller parameters for any feeding device or a feeding unit.

The obtained parameters make it possible to set the required *non-stationary poles* of a closed ACS with full state feedback, *which are determined* by the computer modal control system according to the corresponding algorithm.

6. Conclusion

Thus, the computer modal system developed for controlling the feeding process performs the functions of stabilizing nominal operating modes of the feeding devices unit at the pre-mixing stage, which contributes to the production of high quality mixtures.

Moreover, it should be noted that all necessary procedures associated with the formation of vector-matrix models of the control plant, the assessment of its controllability, and, if necessary, observability, the operations of converting 1D-signals into multidimensional wavelet distributions, as well as operations to implement the matrix algebra used in calculating the parameters of the modal controller, when testing them in real time, it was shown that the required CPU clock frequency of a computer as part of the computer modal control system does not exceed 500 MHz.

Acknowledgements

All comments and suggestions by critical readers and editorial assistance are acknowledged with gratitude.

References

[1] Fedosenkov B A 2005 *Scientific and Engineering Fundamentals of Creating and Modeling the Computer-aided Systems for Controlling Continuous-type Mixture-producing Processes* Abstract of Dr Sci. Dissertation in Automation and Control Systems (Kemerovo) p 55

[2] Mallat S 2005 *Wavelets in Signal Processing* (Moscow: Mir) p 673

[3] Auger F and Chassande-Mottin E 2008 *Quadratic Time-frequency Analysis I: Cohen's Class / in "Time-frequency Analysis: Concepts and Methods"* (ISTE) pp 131–163

[4] Ackermann J E 1972 *Der Entwurf Linearer Regelungs-systeme im Zustandsraum. Regelungstechnik* **20** R 297-300

[5] Strejc V 1985 *State Space Theory of Discrete Linear Control* ed Ya Z Tsypkin (Moscow: NG RF-ML) p 296

[6] Kim D P 2004 *The Automatic Control Theory* vol 2 (Moscow: Fizmatlit) p 464

Structural and parametric non-stationary modal control systems

D B Fedosenkov[1], A A Simikova[2] and B A Fedosenkov[2]

[1]Siberian Generation Company, 53 Dubininskaya Str., build. 5, Moscow, 115054, Russia
[2]T.F. Gorbachev Kuzbass State Technical University, 28 Vesenniaya Str., Kemerovo, 650000, Russia

E-mail : rafwaveletsve@mail.ru

Abstract. The article deals with the aspects of occurring the effect of structural and parametric non-stationary states of various kinds in the models of feeding devices (FD), the latter is considered as the part of a continuous-type mixture-producing aggregate. Relevant features of such an effect are revealed and discussed. In particular, different types of transient modes that occur in the feeding devices under increasing or decreasing the load applied to the intake part of feeders are introduced and described. To demonstrate the complexity of feeding devices models, the distribution of numbers and types of poles for models of different FD-unit configurations have been submitted here. Also we operate with concepts of the so-called autonomously sporadic and forced-modal transient processes, and their images in the forms of 1D-signals and time-frequency distributions are brought in. Finally, the procedures of calculating the modal controller parameters, i.e. the full state feedback matrix, are given in detail.

1. Introduction

When modeling the operating modes of real manufacturing mixture-producing aggregates during multi-ingredient feeding, sporadic changes in the load applied to the intake part of the feeding devices (FD) occur that are caused by feeding into FDs of materials that are initially a priori randomly distributed in the FD-ingredient material containers. As a result, the structural and parametric composition of the control object model can change.

2. The description of structural and parametric non-stationary state

In particular, when the load applied to FD increases, structural adjustment of the feeding process models occurs. For example, when dosing with continuous-type feeding devices (spiral or screw type CTFD [1] – SPFD or SCFD) during load growing (L-rise), a combined transient process (TP) occurs in the form of a flow rate signal at the output of a feeder, including the ordinary TP determined by three poles of the feeding process with a constant feeding frequency, and non-stationary direct time-dependent-frequency TP, in which the nature of the latter is determined by the additional pole creating the exponential form of such a process. Its shape which is well approximated by the impulse response characteristic of the first-order non-periodic process, has been experimentally established (using Cohen's class time-frequency distributions [2]). Thus, the general order of the forward combined TP model becomes equal to four.

With an opposite change in the load on the CTFD, that is with its decreasing (L-dump), the forward non-stationary combined TP arising in the control object is determined – in addition to the influence of three "native" poles on the process – by three more poles that additionally form the forward non-stationary time-dependent-frequency TP. The latter changes according to the two-exponential dependence of the transient characteristic of the second-order non-periodic type. These factors characterize the concept of *a structural non-stationary state.*

IOP Conf. Series: Materials Science and Engineering **865** (2020) 012003 doi:10.1088/1757-899X/865/1/012003

On the other hand, transient processes occurring in the object, are characterized by *a parametric non-stationary state*, which is manifested in the current variability of the instantaneous frequency of the flow rate signal during L-rise / L-dump at the input of a feeding device.

3. Kinds of transient processes and the distribution of poles in the FD-unit

The same kind of transient processes is formed by the FD-computer modal control system [3, 4] during implementing reverse TPs that restore nominal feeding modes. Such transient processes are referred to as upward reverse TPs (URTP) and downward reverse ones (DRTP). Then the forward non-stationary transient processes, respectively, can reasonably be called the downward and upward forward transient processes (DFTP and UFTP).

Thus, under modal control the effect of forced structural and parametric non-stationary state is created due to a change in the localization of the "poles constellation" of the closed automatic control system.

The degree of modal control complexity that is of the closed system with feedback over the vector of state variables [5], is determined by the number of poles in the constellation.

Table 1 shows the distribution of the poles for models of feeding devices composing the feeding devices unit (FDU).

In accordance with the full structure of the FDU including in total five feeders (two continuous-type feeders – SPFD and SCFD, and three portion feeders – PFDs), provided that the latter model has a squeezing pulse index λ of a dose formation other than 2, a block diagram of the control plant model has been formed.

Table 1. The poles available in the models of feeding devices.

FD type	Number of FDs	Number of poles
SPFD	2	- for a single CTFD: 3 (1 pair of conjugated imaginary poles)
SCFD	1	3 (1 pair of conjugated imaginary poles)
CTFD	3	- total number of CTFD poles: 9 (3 pairs of conjugate imaginary poles + 3 real poles)
PFD, $\lambda=2$, N=10	2	- for a single PFD: 11 (5 pairs of conjugated imaginary poles + 1 real pole)
DTFD, $\lambda=2$, N=10	2	- total number of DTFD poles: 22 (10 pairs of conjugated imaginary poles + 2 real poles)
PFD, $\lambda \neq 2$, N=10	2	- for a single PFD: 41 (20 pairs of conjugated imaginary poles + 1 real pole)
PFD, $\lambda \neq 2$, N=10	2	- total number of DTFD poles: 82 (40 pairs of conjugated imaginary poles + 2 real poles)
FDU (2 SPFDs + 1 SCFD + 2 PFDs), $\lambda\neq2$, N=10	5	- total number of FDU poles: 91 (43 pairs of conjugate imaginary poles + 5 real poles)
FDU, (SPFD + SCFD + 3 PFDs), $\lambda\neq2$, N=10	5	- total number of FDU poles: 129 (62 pairs of conjugated imaginary poles + 5 real poles)

4. Autonomously sporadic and forced-modal transient processes

Thus, to initiate the modal control process, information is needed on the previous nature of the feeding process that occurs in the conditions of sporadic changes available in the load on the intake part of the feeding device. We call such primary feeding transient processes under varying load conditions *autonomously sporadic* (they include the downward and upward types of forward transient processes, respectively denoted as DFTP and UFTP), and the transient processes that restore nominal dosing modes and are formed by the modal control system, then are called as *forced-modal* ones (these are the mentioned above URTPs and DRTPs).

The images of autonomously sporadic processes recorded experimentally and displayed in a multidimensional time-frequency format of Cohen's class wavelet distributions [6. 7], and also reverse processes such as URTPs and DRTPs are presented in figure 1 and figure 2.

When the load on the feeder increases, with the help of the wavelet map (a Cohen's class distribution) of the disturbed autonomous-sporadic mode, the system evaluates the time constant T (figure 2), and then relevant required values of time constants T1 and T2 of the upward reverse transient process (URTP) are determined – figure 1.

If the load drops, the modal control system estimates the time constants of the forward process of the UFTP type, according to which a certain time constant of the downward process DRTP is set.

These procedures are performed in real time in the framework of the modal control algorithm, and as a result, the modal controller parameters are calculated, with the help of which reverse transient processes such as URTP and DRTP are implemented.

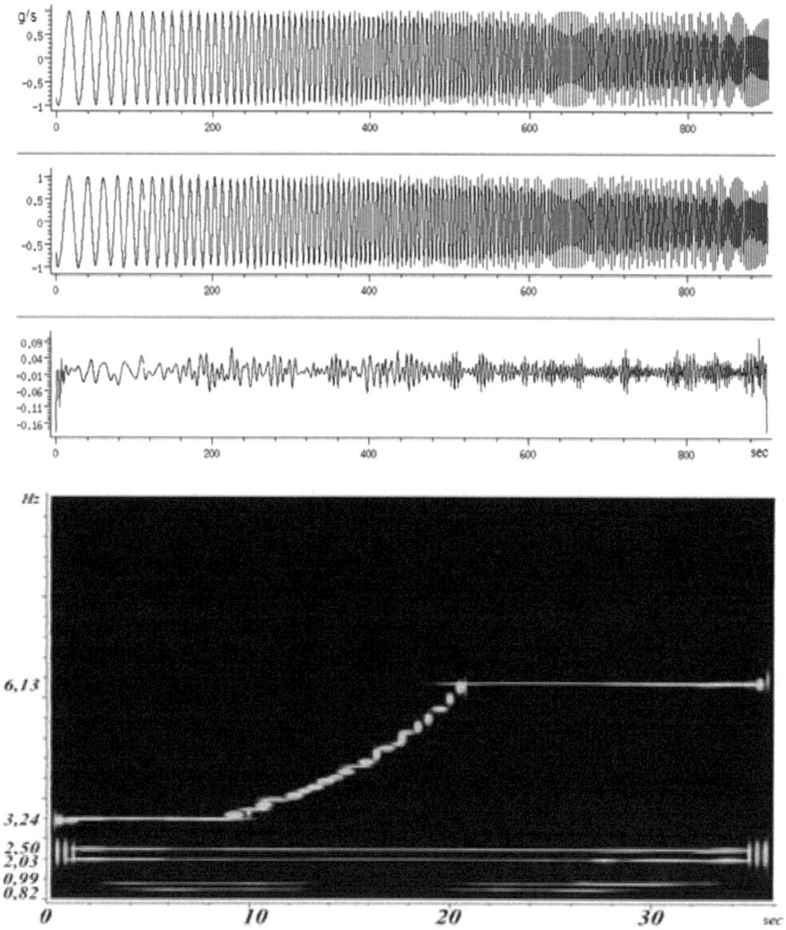

Figure 1. The centered transient process in the spiral feeder in the load reduction mode (chirp-signal); reconstructed by the wavelet matching pursuit algorithm (MP – [2]) feeder flow rate signal, as well as the approximation error of the signal (a); multidimensional chirp-signal image with the increasing time-dependent instantaneous frequency (b).

5. Implementing the algorithm of modal controlling the dosing process

The procedure for implementing the algorithm of modal controlling the dosing process, including the technology of forced localization and relocalization of the closed automatic control system (ACS) poles under conditions of non-stationary processes in the "executive mechanism – FD" system, has been considered on the example of a continuous-type feeder operating in a disturbed mode after a load regime of L-rise type.

It is advisable to apply such a procedure in multidimensional-in-a point systems using the wavelet medium as a means of effective displaying one-dimensional signals. The calculation of the modal controller parameters, that is the determination of the feedback matrix K over the full vector of state variables, is reduced to its calculation by the Ackermann formula [5, 8]:

$$K = \begin{bmatrix} 0 & 0 & 0 & 1 \end{bmatrix} \cdot Q_c^{-1} \cdot \alpha(A)$$

where Q_c^{-1} is the inverse controllability matrix: $\alpha\ (A)$ – the matrix polynomial formed according to the Cayley-Hamilton theorem [5]. To simplify the calculations, the models are taken only for the variable components of the feeding process (the constant components of the process determined by zero poles of feeding process models and non-stationary structural and parametric transient processes are ignored).

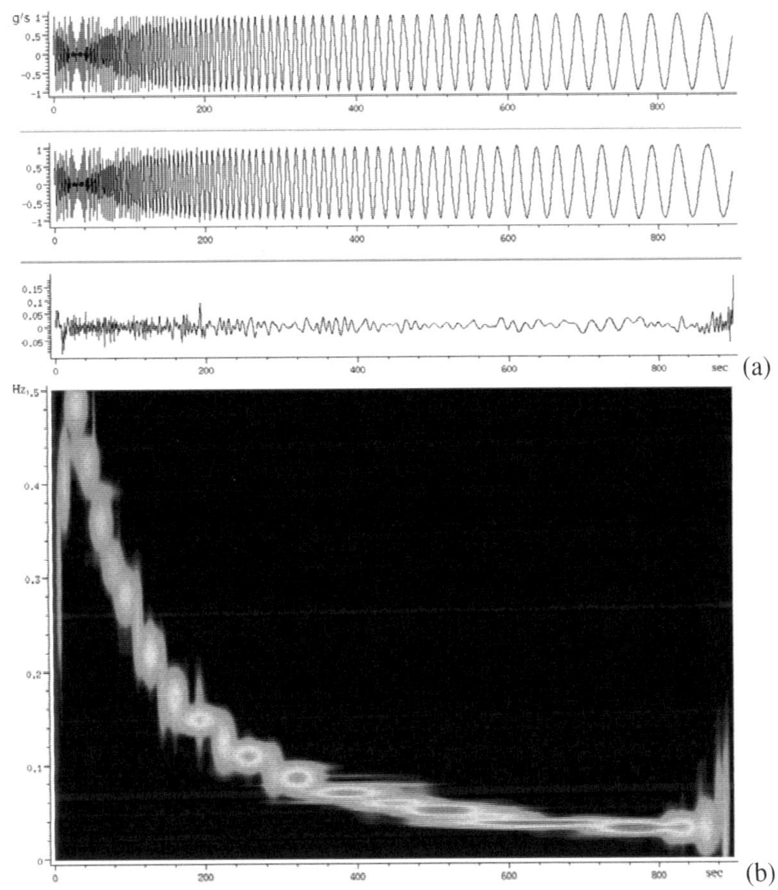

Figure 2. The centered transient process in the screw feeder in the load growth mode (chirp-signal); the signal reconstructed by the wavelet matching pursuit algorithm (MP), as well as the approximation error of the signal (a); multidimensional chirp-signal image with the falling time-dependent instantaneous frequency (b).

The parameters of the inverse controllability matrix, that is the state matrix A and control B, are taken from the vector-matrix model that is implemented within the procedures of the algorithm for modal controlling the feeding process:

$$\mathbf{A} = \begin{vmatrix} A_{11} & A_{12} & A_{13} & A_{14} \\ \dots & \dots & \dots & \dots \\ A_{41} & A_{42} & A_{43} & A_{44} \end{vmatrix}; \qquad B^T = [B_{11} \mid B_{11} \mid B_{11} \mid B_{11}\,]^T$$

$$AB = \begin{vmatrix} (AB)_{11} \\ \dots \\ (AB)_{41} \end{vmatrix},$$

where $(AB)_{11} = A_{11}B_{11}+\ldots+A_{14}B_{41}$;

......

$(AB)_{41} = A_{41}B_{11}+\ldots+A_{44}B_{41}$.

$$A^2B = A(AB) = [(A^2B)_{11}|(A^2B)_{21}|(A^2B)_{31}|(A^2B)_{41}]^T,$$

where $(A^2B)_{11} = A_{11}(AB)_{11}+\ldots+A_{14}(AB)_{41}$;

....

$(A^2B)_{41} = A_{41}(AB)_{11}+\ldots+A_{44}(AB)_{41}$

$$A^3B = A(A^2B) = [(A^3B)_{11}|(A^3B)_{21}|(A^3B)_{31}|(A^3B)_{41}]^T,$$

where $(A^3B)_{11} = A_{11}(A^2B)_{11}+\ldots+A_{14}(A^2B)_{41}$;

....

$(A^3B)_{41} = A_{41}(A^2B)_{11}+\ldots+A_{44}(A^2B)_{41}$.

Thus, the controllability matrix:

$$Q_c = [B|(AB)_{11}|\ldots|(AB)_{41}]^T|[(A^2B)_{11}|\ldots|(A^2B)_{41}]^T|[(A^3B)_{11}\ldots|(A^3B)_{41}]^T],$$

and the inverse controllability matrix is as follows:

$$Q_c^{-1} = Adj[Alg(Q_c^T)] / detQ_c.$$

The desired characteristic polynomial is determined by the formula $\alpha(s) = s^4 - d_1 s^3 + d_2 s^2 - d_3 s + d_4$, where $d1 = (s_{p1} + s_{p2} + s_{p3} + s_{p4})$; where s_{pj}– object poles; $d_2 = [s_{p1} s_{p2} s_{p3} (s_{p1} + s_{p2}) + s_{p4} (s_{p1} + s_{p2} + s_{p3} + s_{p4})]$; $d_3 = [s_{p1} s_{p2} s_{p3} + s_{p1} s_{p2} s_{p3} s_{p4} (s_{p1} + s_{p2})]$; $d_4 = s_{p1} s_{p2} s_{p3} s_{p4}$.

Consequently, the matrix polynomial has the form $\boldsymbol{\alpha(A)} = \boldsymbol{A}^4 - d_1 \boldsymbol{A}^3 + d_2\boldsymbol{A}^2 - d_3\boldsymbol{A} + d_4\boldsymbol{I}$,
$\boldsymbol{I}$ – the unity matrix.

$$\alpha(A) = \begin{bmatrix} 0 & -T^{-1}\omega^2+1 & -\omega^2+1 & -T^{-1}+1 \\ 0 & -\omega^2(T^{-2}+T^{-1}) & -\omega^2+1 & (-\omega^2+1)+(T^{-2}-T^{-1}) \\ 0 & \omega^4T^{-1}-\omega^2(T^{-2}+T^{-1}) & \omega^2(\omega^2-1) & \omega^2(T^{-1}-1)-T^{-3}+(T^{-2}-T^{-1})+1 \\ 0 & \omega^2(\omega^2-1)(T^{-2}+T^{-1}) & \omega^2(\omega^2-1) & \omega^2(\omega^2-T^{-2})-\omega^2(T^{-1}-1)+T^{-4}-T^{-3}+(T^{-2}-T^{-1}) \end{bmatrix}$$

Looking at $\alpha(A)$, it is easy to see that in the open ACS the poles s_{p2}, s_{p3}, and s_{p4} (two imaginary conjugate and the real negative) corresponding to the parameters ω and T, affect the transient process type, and just those ones create the frequency component and determine the speed of the transient process.

As a result, we obtain the modal controller matrix K:

$$\boldsymbol{K} = [0\ \ 0\ \ 0\ \ 1]\,\boldsymbol{Q_c}^{-1}\,\boldsymbol{\alpha(A)} = [0\ \ |(T^1\omega^2-1)\ \ |(\omega^2-1)\ \ |(T^1-1)] = [K_1\ \ |K_2\ \ |K_3\ \ |K_4].$$

This matrix is continuously recalculated in real time with an arbitrary step of the millisecond range during the transient process to ensure its time-frequency (chirp-) nature.

Using similar procedures, the computer control system calculates the modal controller parameters for any feeding device or a feeding devices unit.

6. Conclusion

The obtained modal controller parameters make it possible to set the required non-stationary poles of the closed ACS with full state feedback, which are determined by the computer modal control system according to the developed algorithm.

Thus, it was revealed that when modeling the processes in the computer system of modal controlling the feeding regimes, non-stationary structural and parametric transient processes occur, that are caused by secondary reactions to sporadic changes in the input uncontrolled influences (load variation) applied to the intake parts of feeding devices. Models of this kind are the constituent elements of the combined algorithm for modal controlling technological processes of various natures.

Acknowledgements

All comments and suggestions by critical readers and editorial assistance is greatly appreciated.

References

[1] Fedosenkov B A and Fedosenkov D B 2015 *The Computer-aided Controlling the Mixture-Producing Processes in a Wavelet Medium* (Kemerovo: Kemerovo Institute of Food Science and Technology) p 189

[2] Mallat S 2005 *Wavelets in Signal Processing* (Moscow: Mir) p 673

[3] Simikova A A, Fedosenkov D B and Fedosenkov B A 2018 *Proc. of the 2nd Int. Conf. on Innovations in Information Technologies, Mechanical Engineering, and Motor Transport* (Kemerovo: KuzSTU) pp 136–9

[4] Fedosenkov D B, Simikova A A and Fedosenkov B A 2019 *Proc. of Int. Conf. on Innovation Aspects of Development in Engineering Technologies* (St. Petersburg: SPbF SIC MS) vol 2 pp 162–7

[5] Kim D P 2004 *The Automatic Control Theory* vol 2 (Moscow: Fizmatlit) p 464

[6] Cohen L 1995 *Time-frequency analysis* (Englewood Cliffs: Prentice Hall) p 299

[7] Auger F and Chassande-Mottin E 2008 *Quadratic time-frequency analysis I: Cohen's Class / in "Time-frequency Analysis: Concepts and Methods"* (ISTE) pp 131–63

[8] Ackermann J E 1972 *Der Entwurf Linearer Regelungs-systeme im Zustandsraum.– Regelungstechnik* 20 R 297–300

Development and creation of an innovative scientific and educational cluster for the integrated use of coal and its products

V I Murko[1], E V Protopopov[1], M V Temlyantsev[1], V V Chaplyigin[1], Yu A Litvinov[1] and M A Volkov[2]

[1]Siberian State Industrial University, 42 Kirova Street, Novokuznetsk, 654007, Russia
[2]LLC SibNIIugleobogashcheniye, branch in Prokopyevsk, 1 Gornaya Street, Prokopyevsk, 653000, Russia

E-mail: sib_eco@mail.ru

Abstract. The article proposes to implement a cluster approach to training, development and creation of high-tech industries in the field of processing of coal raw materials with obtaining end products with high added value. The main technological processes used are: preparation and burning of coal water slurry, based the finely dispersed coal preparation waste as well, liquefaction of demineralized coal and coal slime using a complex of high-energy influences (cavitation, ultrasonic and electric pulse), processing of waste from coal processing plants, bottom ash wastes as a basis for filling mixtures in the dead underground workings and open-pit mines. The implementation of the proposed cluster approach will allow the current scientific and practical problems facing the region in the field of coal mining and processing to be solved.

1. Introduction

One of the effective methods for increasing the rate of knowledge diffusion and training with the concentration of high-tech industries is the organization and creation of innovative clusters. The result is a synergistic effect of technological, technical, economic and institutional factors.

For Kuzbass, where one of the main sectors of the national economy is the coal industry, the rational use of mined coal and its products is one of the main tasks for the successful development of the region. Considering the significant transport costs when delivering coal to its customers both to the West and the East, coal companies have put a lot of effort and invested heavily in the construction of new coal processing plants (CPP). According to statistics, almost all coal of coking grades is currently being processed in Russia (in 2014, 94.2 million tonnes were extracted – 90.0 million tonnes were processed per year), respectively, for power-generating coal these figures are 345.1 million tonnes per year and 110.0 million tonnes per year.

The processing technology at CPPs is characterized by the use of closed water slurry schemes to produce coal concentrates with the required moisture content without thermal drying of small classes of coal. The products at CPPs are: concentrate with an ash content of 5.5-10%, industrial product with an ash content of 25-38% (usually for coking coal grades) and a rock with an ash content of more than 70%.

At the same time, finely dispersed coal preparation waste (FDCPW) of the 0-0.5 (1) mm class, usually represented by filter-press residual matter – filter cake (FC), appeared in the composition of the rock being shipped to the dump. The output of the filter cake is from 5 to 10% by weight of the processed coal at the CPP. Given the high humidity and ash content of this product, its implementation on the market without preliminary processing is not economically feasible.

A feature of the CPP filter cake is also that this product contains extremely toxic flocculants, which entails environmental pollution around their storage sites. As a result, coal enterprises suffer economic losses due to the need to use a significant amount of vehicles (the specific cost of transporting the filter cake is about 10 rubles per tonne·km, as well as paying the cost of waste storage, and the surrounding territories get serious environmental problems.

In addition, significant environmental problems arise in settlements located near central heating and power plants (CHPP) and boiler houses associated with the storage of huge quantities of bottom ash wastes (BAW) that remain in ash dumps and residual tanks for decades. As a result, such a notion as "dirty coal" appeared in the mass media and political space, the use of which as a technological and in fuel raw materials must be reduced, and if possible, completely abandoned. At the same time, possible social, economic and political consequences are not taken into account, especially for regions such as Kuzbass.

Thus, the most urgent tasks in the extraction, processing and use of coal and its products, in our opinion, are:

- processing and efficient use of finely dispersed coal preparation waste and low-grade coal and industrial product;
- processing and use of the BAW by CHPPs and boiler facilities;
- reclamation of disturbed lands during the operation of coal mines and open-pit mines.

2. Results and discussion

To solve the set tasks it is proposed to use a cluster approach, when on the same platform it is proposed to combine both modern innovative technologies, which allow individual technical problems of processing and using coal and its products to be solved, including deep processing, and to gain new knowledge in their study and implementation.

Thus, an innovative scientific and educational cluster for the integrated use of coal and its products, has been developed and is being implemented at SibSIU, including:

- processing of coal and coal slimes;
- technology and equipment for coal processing, including its crushing and classification;
- technology and equipment for the preparation of slurry water-coal fuel (SWCF) based on coal, industrial product, and FDCPW;
- technology and equipment for hydrotransportation of the obtained "liquid" coal through the pipeline and its storage in containers;
- technology and equipment for producing briquette fuel based on coal and FDCPW;
- technology and equipment for producing coal oil granulate and, on its basis, coal-water-oil suspension and coal oil briquettes;
- technology and equipment for burning the obtained types of fuel in a boiler plant with a fuel chamber of layer and swirling-type furnace;
- bench installation for deep processing of coal-water-oil suspension to obtain a synthetic oil product;
- technology and equipment for producing a hardening filling on the basis of CPP and BAW rock, ash from CHPP, also from boiler houses, including those obtained during the combustion of SWCF;
- technology and equipment for producing slurry based on rock and BAW for filling the mined territories during their reclamation;

- a modern coal-chemical laboratory with a collective use center, where all types of analysis of coal and their products are carried out.

It is the implementation of the proposed integrated cluster approach that allows ambitious scientific and practical tasks to be set, a search for solutions to be offered to students and graduate students, employees of design organizations and industrial enterprises, and representatives of the business community. At the same time, the creation of a demonstration technological complex in the block of heavy laboratories of SibSIU within the framework of the innovation cluster will allow the specific narrow technical problems in coal processing and use of the obtained products to be solved with achievement of the required characteristics and environmental effect.

In recent years, a significant number of publications have appeared that describe technologies for the utilization of FC by burning it in boilers with a boiling bed and a swirling combustion system. The main technologies that allow economically efficient and environmentally safe disposal of FDCPW are technologies for producing fuel briquettes and SWCF.

However, at the moment, these technologies are not widely used in Russia due to the following reasons:

- the presence on the market of Kuzbass and other coal regions of a significant amount of relatively cheap raw coal, and transportation of briquettes with high ash content (up to 35-40%) over considerable distances (even to nearby regions) is unprofitable;
- the lack of demand for cheap SWCF from local industrial and municipal enterprises that have boiler houses, CHPPs due to the small experience in the industrial operation of facilities using this technology, including due to significant fluctuations in the quality of FDCPW (ash, humidity);
- the presence of the subjective factor "we will not be the first", which is commonly found in the management of modern coal and energy companies;
- lack of a legislative framework for the use of coal preparation wastes and BAW in terms of the reliability of long-term forecasts of their value.

Thus, the search and development of technology and a set of equipment for the efficient use of coal and its products is an urgent problem for coal companies both in terms of economics and environment. Taking into account the significant variation in the quality of coal and its products by size, the yield of volatiles (coal grade), moisture and ash content, in order to develop optimal technological solutions for their use, it was decided to create a modern experimental base. A technological complex consisting of four main modules was created at Siberian State Industrial University (SibSIU) together with Scientific and Production Center "Sibekotekhnika", Scientific and Production Center "Sinoyl" and the branch of Siberian Research Institute of Coal Processing in Prokopyevsk: coal preparation and preparation of fuel, transportation, storage of fuel, and burning of different types of fuel and liquefaction of coal (figure 1).

The coal preparation and processing module with the preparation of various types of fuel contains: a hammer mill, a batch auger mixer, a coarse vibration filter, a drum mini-bead beater and rod mill of discontinuous operation, a vibratory mill of discontinuous and continuous operation, pumping equipment, and tanks for finished fuel storage. The module also has equipment for processing the feedstock using oil granulation and producing briquettes from various feedstock. Thus, the set of mounted equipment for fuel preparation provides the possibility for implementation of various methods for its preparation, based on the type of feedstock (raw coal, coarse slime, FDCPW, coal-oil granules). The fuel transportation and storage module includes: a fuel storage tank, a transportation pump of volumetric type, a pipeline consisting of horizontal, vertical and inclined sections, start-up chambers and reception of calibrating scrapers, instrumentation and automation devices (flow meter, pressure gauges, flow viscometer). The created module allows hydrotransportation studies of various types of water-coal suspensions (coarse and finely dispersed, low and highly concentrated) in turbulent, transitional and laminar modes to be conducted. In this case, the specific pressure losses, the presence of residual matter in the pipeline, the actual rheological characteristics necessary for the calculation of pipeline systems are determined.

IOP Conf. Series: Materials Science and Engineering **865** (2020) 012004 doi:10.1088/1757-899X/865/1/012004

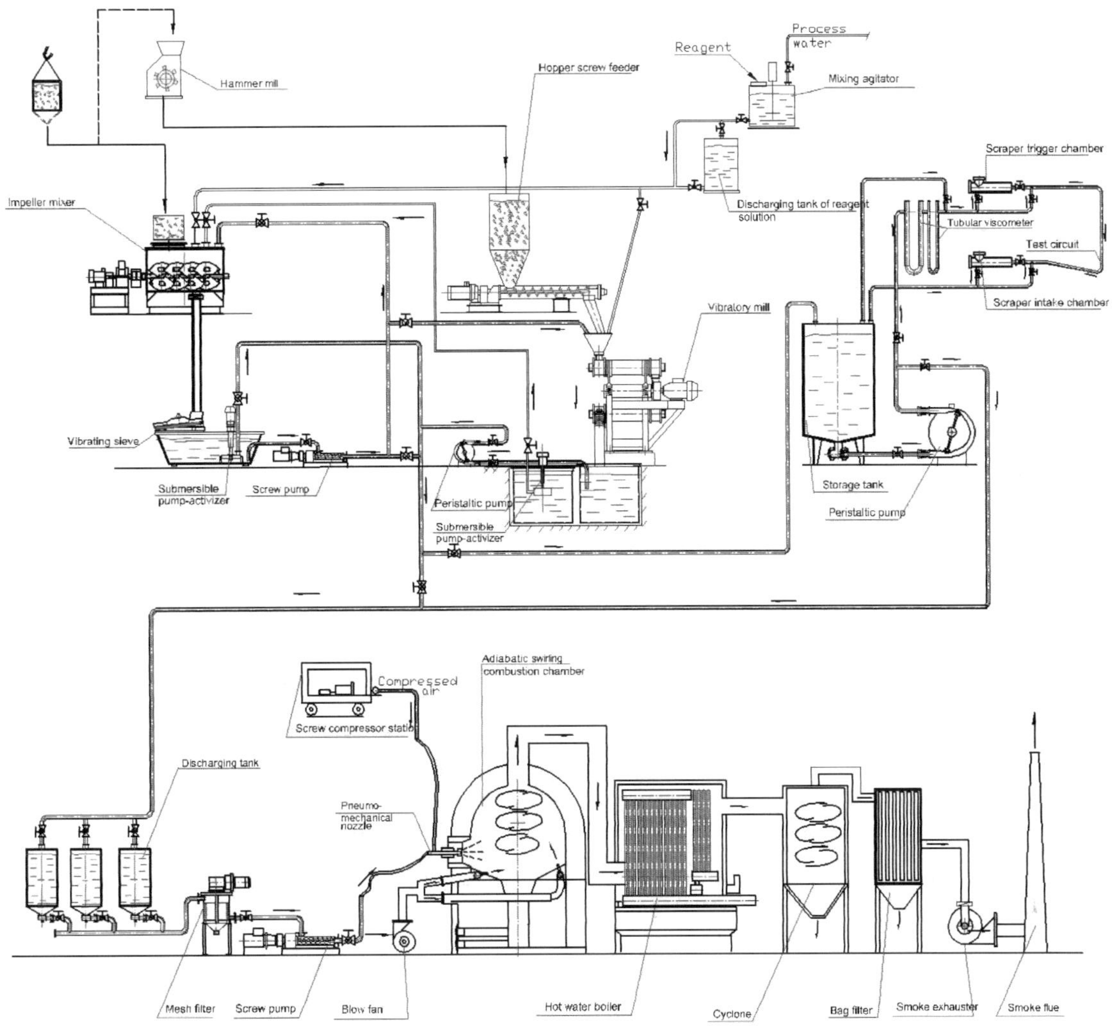

Figure 1. Technological scheme of the bench installation.

The combustion module contains: a fuel combustion chamber with a bed and swirling combustion systems, a hot water boiler with a heat power of 0.25 MW, a two-stage dust collection system, a chimney, draft equipment, a metering pump for injecting fuel into the combustion chamber, a compressor with a receiver for supplying compressed air to pneumomechanical fuel injector, instrumentation kit. To ensure year-round operation of the bench installation, a boiler water cooling system is provided in the air cooling apparatus. The operation of the combustion module makes it possible to determine the efficiency of fuel combustion, the efficiency of the boiler on various types of fuel, the composition of the flue gases.

The operation of the bench installation is carried out both remotely and locally. In addition, an automation system was mounted by the processes of preparation and fuel burning, which allows work of creating a technological complex in automatic mode to be performed.

Alongside with a bench installation, a coal chemistry laboratory was also created.

Testing of the bench installation operation was carried out using FDCPW at CPP "Shchedrukhinskaya" (JSC "Topprom", Novokuznetsk).

Testing of the bench installation showed the wide possibilities of its use for working out the modes for preparing SWCF from feedstock of various quality, both by size and ash content, as well as burning the resulting fuel with the measurement of the necessary technological and technical parameters. When using a gas analyzer, the level of harmful emissions in flue gases is measured.

Thus, the created bench installation allows the technology for the utilization of finely dispersed coal-enrichment waste to be fully implemented by preparing fuel based on it and then burning it. BAW obtained after burning SWCF at CHPPs and rock from CPP are used to prepare hardening and filling mixtures for filling the mined territories and open-pit mines.

A fundamentally new technological process for processing coal and coal slimes is a method for producing artificial "coal" oil. Coal liquefaction provides an alternative low-cost source of oil and oil products. Liquefaction of coal can also be used as a way of waste recycling in coal production and coal processing. The use of coal for the production of synthetic oil is advisable because of its close chemical composition of natural raw materials. The hydrogen content in oil is 15%, in coal – 8%, and carbon – comparable amounts.

SWCF is a finished product for the use in the electric pulse method of liquefying coal. An important indicator in this case is the coal ash content in SWCF, since during hydrogenation the remainder of non-reactive resin compounds mixed with coal and ash remains. To reduce this residue and increase the conversion of coal, i.e. to reduce the loss of carbon, it is necessary to have the lowest possible ash flow into the process. The method of oil granulation is used for this purpose.

Coal is a poly-conjugated system, mainly of a non-aromatic nature, which includes high molecular weight and low molecular weight organic substances bound by intermolecular interaction (IMI) of different nature and strength, and electrons and radicals, interactions with a carbon substance, contribute to the destruction of IMI and hydrogenation [7]. The objective of the liquefaction process is to carry out the depolymerization reaction of the substance and hydrogenation of the products of depolymerization. These reactions are initiated by special high-voltage pulsed discharges in a mixture of coal, water and an organic solvent. In the discharge channel, the substance heats up to hundreds of thousands of degrees and, accordingly, expands sharply, creating high pressure up to several thousand MPa, causing, among other things, cavitation phenomena. Appearing particles of radicals *O, *H, *OH and free electrons activate the suspension as a whole and contribute to the destruction of long molecules of substances, leading to the appearance of liquid organic products. Atomic hydrogen is involved in the hydrogenation reaction, closing newly formed molecular compounds. In addition, during a high-voltage discharge, additional grinding of coal particles occurs due to tensile mechanical stresses – the product decompresses, i.e. additional pores appear that increase solvent access to coal particles [8].

Water facilitates the occurrence of a high-voltage discharge and is a source of *O, *H, *OH radicals, and hydrocarbons are a supplier of hydrogen. The presence of water and hydrocarbons is necessary as one of the components of a supercritical fluid that is formed locally during cavitation. The process of liquefaction in a supercritical environment occurs best in the presence of organic hydrocarbons and water.

Implementation of the proposed method of liquefaction is carried out on a specially created bench installation of electropulse liquefaction.

3. Conclusion

The developed and created innovative scientific and educational cluster allows the modern approach to training, conducting scientific research, introduction of high-tech industries in the field of processing and use of products of coal enterprises to the market to be fully implemented.

Acknowledgments

The study was carried out with the financial support of the Russian Foundation for Basic Research in the framework of scientific project No. 20-43-420016/20.

References

[1] Gokhberg L M, Klepach A N et al 2016 *Methodological Materials for the Development, Implementation of Programs for the Development of Innovative Territorial Clusters and Regional Cluster Policy*
[2] Bortnik I M, Gokhberg L M et al 2015 *Cluster Policy: Concentration of Potential for Achieving Global Competitiveness* (St. Petersburg: Corvus)
[3] Kozlov V A 2017 *Coal* **8** 140–145
[4] Murko V I, Khyamyalyainen V A, Petrova B M 2018 *Proc. of the Conf. "Coal in the 21st Century: Mining, Intelligent Equipment and Environment Protection"* pp 337-340
[5] Murko V I, Ismagilov Z R et al 2017 *Coal Chemistry and Ecology of Kuzbass. Collection of Abstracts* 40
[6] Murko V I 2016 *Proc.of the IX Int. Sci. Conf. in Innovations in Technology and Education* pp 189–190
[7] Rusyanova N D 2003 Coal Chemistry (Moscow: Nauka)
[8] Kurets V I, Usov A F and Zuckerman V A 2002 *Electropulse Disintegration of Materials* (Apatity: Kola Science Center RAS)

On the development of a base of normative situational clock periods of the rational rolling complex functioning

S M Kulakov, A I Musatova and V N Kadykov

Siberian State Industrial University, 42 Kirova Street, Novokuznetsk, 654007, Russia

E-mail: musatova-ai@yandex.ru

Abstract. The main idea of the article is to solve the problem of assessing the normative values of situational clock periods of the rolling complex equipment. The result of the study is a measure of clock cycles, structured by the type of product, technological routes and production situations. An example of rationalization of one of the technological routes of a specific rolling production using situational and normative clock periods is shown, which allows the main technical and economic indicators of the complex to be improved. A situational clock period approach is proposed to build a base of normative indicators for the rolling complex. The following areas of application of the base of indicators are identified: optimization of technological routes, scheduling of the complex, operational planning of the production.

1. Introduction

The tasks of scheduling and operational management of a complex production require the preliminary development of a base of situational and normative indicators of its functioning, in which clock periods are the determining indicators. The paper presents the synthesis of the procedure for assessing the standard values of clock periods taking into account the amount of equipment used and the product mix. The object of the study is a rolling complex, which includes a billet (breakdown) shop and a long product rolling mill, consisting of four parallel processing mills (medium-section and three small-section mills). The connecting link in this complex is an intermediate warehouse of blanks of the breakdown workshop, designed to maintain a rational stock of blanks with the aim of uninterrupted supply of rolling department with metal.

2. Characterization of technological routes of the rolling complex

Currently, the material flows of the studied rolling complex are organized according to six routes (figure 1).

The considered main technological routes No. 1, No. 2 are organized in the system "breakdown mill → refrigerator → billet warehouse → rolling mills", the so-called "cold charging" of billets coming from the warehouse by the roller conveyor to the heating furnaces of four section rolling mills. In addition, a transit route No. 3 "breakdown mill → mill 450" was organized, providing for the so-called "hot charging" of workpieces coming from 400 tonne cutters through the system of transmitting devices (slippers, roller conveyor) in hot condition immediately to the heating furnaces of mill 450, which greatly saves time and fuel resources. Breakdown bars intended for the production of semi-finished products (blooms and slabs) after cleaning surface defects are fed to 1400 t cutters, where they are cut to appropriate lengths, according to customer orders, then the semi-finished products are transported to the storage area for rolled metal products (routes No. 4, No. 5) .

Part of the metal enters the rolling complex from the continuous steel casting department of the oxygen converter shop in the form of cast billets with a section of 150×150 mm, which are transported directly to the billet warehouse, from where they are subsequently fed to the heating furnaces of mill 450 (route No. 6).

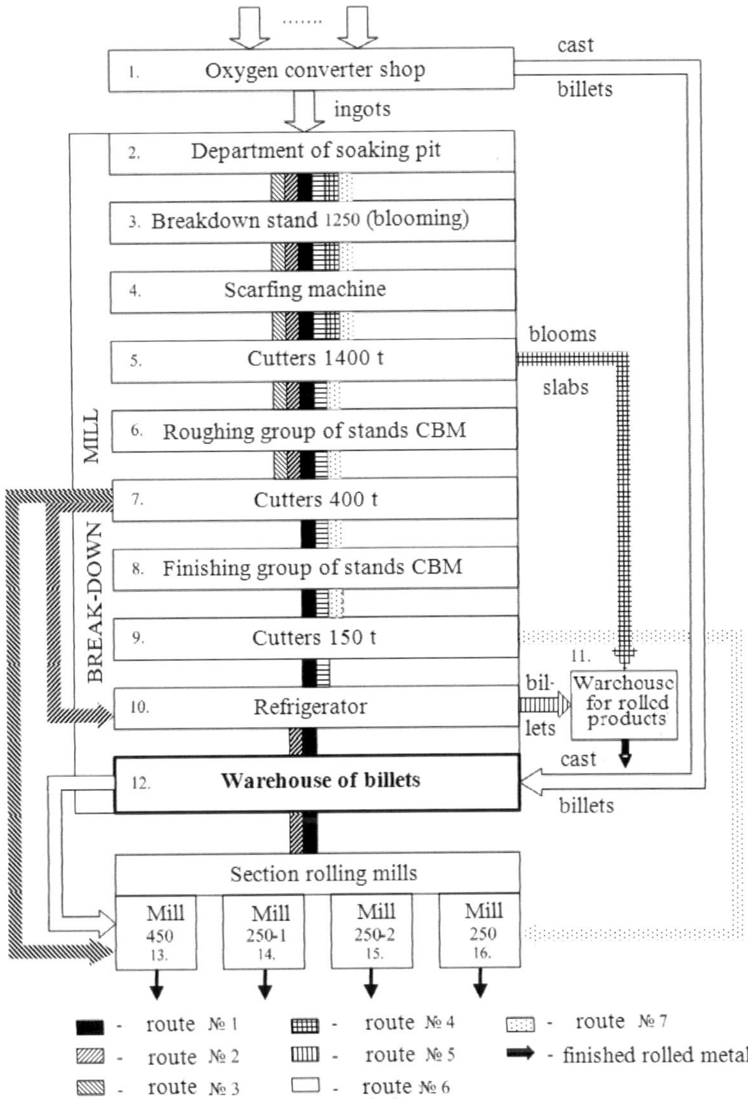

Figure 1. Scheme of technological routes of the rolling complex.

3. Stages of development of situational-clock models

The methodology for constructing interconnected technically feasible and normative cycles of equipment operation, their classification and application in rolling production are considered in [1, 2].

In order to identify the internal reserves of production, improve the organization, planning and operational management of material flows, a multivariate assessment of the measures was performed for situations characterizing the functioning of the investigated rolling complex on the basis of the developed situational clock models in the following stages.

1. The definition of technically possible T^{tp} and normative T^{n} clock periods of the main and auxiliary equipment of the complex:

- in the breakdown mill (clock periods of the processes: heating in the department of soaking pits, rolling at blooming and at the CBM, peeling at the scarfing machine, cutting on cutters and cooling on sections of the refrigerator), $T^{tp(n)}_{tot(wm)}$;
- on a medium-section mill 450 (clock periods of processes: heating in walking hearth furnaces, continuous rolling in stands, cooling in the refrigerator, dressing on straightening machines, cutting by cutters and baling), $T^{tp(n)}_{450}$;
- on small-section mills 250-1 and 250-2 (clock periods of processes: heating in continuous furnaces, continuous rolling into two billets in stands, cooling in the refrigerator, cutting by cutters scissors and baling), $T^{tp(n)}_{250\text{-}1}$, $T^{tp(n)}_{250\text{-}2}$;
- at a wire mill 250 (clock periods of processes: heating in continuous furnaces, continuous rolling into four billets in stands, forming coils in special machines, cooling of coils on a hook conveyor and their strapping), $T^{tp(n)}_{250}$.

2. The selection of the technically possible (normative) clock period of the *k*-th unit of the complex, including *n* of series-connected units of equipment, was carried out in accordance with the rule:

$$T^{tp(n)}_{k} = max\left\{T^{tp(n)}_{k.1}, T^{tp(n)}_{k.2}, \dots T^{tp(n)}_{k.n}\right\}, \tag{1}$$

where $T^{tp(n)}_{k.n}$ is the clock period of the *n*-th equipment of the *k*-th unit, s.

3. Determination of the clock period of the rolling complex according to the *z*-th technological route, taking into account production situations during the joint work of *k*-th units $T^{tp(n)}_{z}$.

The possibility of introducing an additional transit route No. 7 of hot blanks on the line "breakdown workshop – wire mill 250" was revealed in order to reduce the consumption of fuel and energy resources on the basis of studying the existing organization of material flows in the rolling complex, plans for the location of technological and transport equipment in the billets warehouse, as well as the obtained estimates of the clock periods of the breakdown workshop and section rolling mills. In this case, breakdown bars with a cross section of 100×100 mm after cutting them by cutters of 150 tonnes, without cooling in the refrigerator, are transferred by a roller conveyor (which must be installed in the warehouse) to the feed roller to the continuous furnace No. 2 of the wire mill. It should be noted that the transit for furnace No. 1 of mill 250 is not provided, furnace No. 1 is located at a great distance from refrigerators, which will lower the temperature of the transit metal during its transportation to this furnace.

The work of the rolling complex can be carried out according to the options that correspond to the following production situations (*Sit*), figure 2, depending on the type and size of the blanks (numbers correspond to the blocks of figure 1).

The clock periods of the breakdown mill and section rolling mills were simulated for the reduced unit of production (piece) for which an ingot of a certain type was adopted, comparable in weight to a package of workpieces of a given size (by cross-section and length). In this case, possible situations were taken into account, each of which is characterized by technological modes of heating, rolling and cooling, the amount of auxiliary equipment used (heating and cooling devices, cutting and baling units), the combination and direction of material flows.

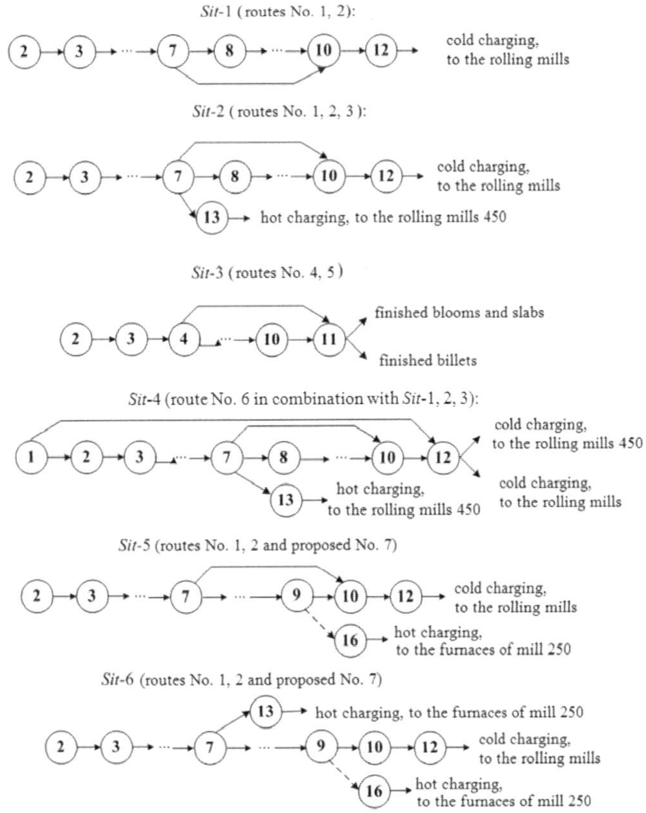

Figure 2. Scheme of production situations according to the technological routes.

4. Modeling and selection of rational technological routes

Multivariate modeling allowed the rational conditions for the implementation of technological routes to be determined. Fragments of the results of modeling and evaluation of the clock cycle of the rolling complex are presented for some situations in tables 1-3. Note that the tables show (as characteristic examples) only one of the many options for the developed regulatory database, taking into account combinations of possible modes, the amount of equipment used and product characteristics.

For the trouble-free operation of the rolling mills during the cold charging of the blanks (*Sit*-1), the clock period of the breakdown mill $T_{tot}^{tp(n)}$ should be less than the clock period of the rolling mills (equivalent clock period) $T_{eq}^{tp(n)}$ that is, $T_{tot}^{tp(n)} < T_{eq}^{tp(n)}$

In the inequality, the less than sign indicates that the breakdown mill provides scrap metal not only for the rolling mills, but also produces additional metal for the sale of commercial products (blooms, slabs, billets) and for creation, if necessary, of current and technological stock of blanks in the warehouse.

$$T_{eq}^{tp(n)} = T_{250\text{-}1}^{tp(n)} \cdot \gamma_{250\text{-}1} + T_{450}^{tp(n)} \cdot \gamma_{450} + T_{250}^{tp(n)} \cdot \gamma_{250} + T_{250\text{-}2}^{tp(n)} \cdot \gamma_{250\text{-}2} \quad (2)$$

where $\gamma_{250\text{-}1}$, γ_{450}, γ_{250}, $\gamma_{250\text{-}2}$ – the number of bundles of blanks corresponding to the clock periods of the rolling mills, pcs.

IOP Conf. Series: Materials Science and Engineering **865** (2020) 012005 doi:10.1088/1757-899X/865/1/012005

During the operation of the heating furnaces of mill 450 with cold charging of the billets, the breakdown mill can work in two directions, that is, transfer bundles of blanks in transit to the furnaces of mill 450 and to the warehouse into stock heaps (*Sit*-2) in the following ratio:

$$T_{450}^{tp(n)} \geq T_{tot(450)}^{tp(n)} \cdot \gamma_{tot(450)} + T_{tot(wm)}^{tp(n)} \cdot \gamma_{tot(wm)}, \quad (3)$$

where $T_{tot(450)}^{tp(n)}$, $T_{tot(wm)}^{tp(n)}$ are the clock periods of the breakdown mill during the hot transit of blank bundles to the furnaces of mill 450 and storing the bundles into stock heaps, s; $\gamma_{tot(wm)}$, $\gamma_{tot(450)}$ – the number of bundles of blanks supplied from the breakdown mill to the furnaces of mill 450 and to the warehouse, pcs.

For example, the clock period of the breakdown mill during ingot rolling $T_{tot(wm)}^{tp}$ = 42.55 s; the clock period of mill 450 when rolling a bundle of billets into finished section (channel steel) is T_{450}^{tp} = 128.82 *s*, then according to formula (1):

$T_{tot(wm)}^{tp} = max\{40.76; 38.75; 27.72; \mathbf{42.55}; 30.05; 38.94\} = \mathbf{42.55}$ *s* (per bundle of blanks)

$T_{450}^{tp} = max\{10.07; \mathbf{21.47}; 18.9; 18.3; 17.4\} = \mathbf{21.47}$ *s* (per blanks) or 128.82 c (per bundle).

Consequently, the breakdown mill must ensure the supply of a bundle of hot billets to mill 450 every 128.82 s, while it can submit two more bundles to the warehouse for further transfer to small-section mills, table 1.

Table 1. Estimation of the clock period of the rolling complex for *Sit*-2 (example).

Clock period				Clock supply of bundles (γ = 2) to the warehouse	The clock period of the complex, taking into account the number of bundles (γ) supplied to	
Breakdown mill	Mill 450	Fridge			mill 450	warehouse
	taking into account the work of three furnaces	number of sections				
		9	10		one	two
Technically possible clock periods, s						
42.55	128.82	38.94	35.05	85.1	128.82	
Normative clock periods, s						
47.28	140.04	44.25	39.83	94.56	140.04	

When operating a heating furnace No. 2 of wire mill 250 with a hot charge of billets (*Sit*-5), the following condition for the rational functioning of the complex must be observed (table 2):

$$T_{250}^{tp(n)} \geq T_{tot(250)}^{tp(n)} \cdot \gamma_{tot(250)} + T_{tot(wm)}^{tp(n)} \cdot \gamma_{tot(wm)} \quad (4)$$

The functioning of the rolling complex during the operation of the heating furnaces of mill 450 and the furnace of mill 250 with a hot charge of workpieces (*Sit*-6) should be carried out in the ratio:

$$T_{250}^{tp(n)} \geq T_{tot(250)}^{tp(n)} \cdot \gamma_{tot(250)} + T_{tot(450)}^{tp(n)} \cdot \gamma_{tot(450)} + T_{tot(wm)}^{tp(n)} \cdot \gamma_{tot(wm)} \quad (5)$$

Table 2. Estimation of the clock period of the rolling complex for *Sit*-5 (example).

Breakdown mill	Clock period (per ingot)		Clock supply of bundles (γ = 11) to the warehouse	The clock period of the complex, taking into account the number of bundles (γ) supplied to	
	Mill 250	Fridge		mill 250	warehouse
	taking into account the work of one furnace	number of sections 10		one	eleven
Technically possible clock periods, s					
42.55	510.9	31.25	468.05	510.9	
Normative clock periods, s					
47.28	601.12	35.51	520.05	601.12	

Table 3. Estimation of the clock period of the rolling complex for *Sit*-6 (example).

Breakdown mill	Clock period (per ingot)				Clock supply of bundles (γ = 7) to the warehouse	The clock period of the complex, taking into account the number of bundles (γ) supplied to		
	Mill 250	Mill 450	Fridge			mill 250	mill 450	warehouse
	taking into account the work furnaces		number of sections					
	one	three	8	9		one	four	seven
Technically possible clock periods, s								
42.55	510.9	128.82	38.6	34.5	297.85	510.9		
Normative clock periods, s								
47.28	601.12	140.04	43.86	39.2	330.94	601.12		

5. Conclusion

Thus, the introduction of an additional transit technological line to the wire mill will make it possible, with small capital investments, to reduce the current costs of the wire mill by reducing the heating time and fuel supply rate, since one furnace will work in the preheating mode (*Sit*-5, 6).

Implementation of the proposed organizational and technical measure (project route No. 7) from the 150 tonnes cutters of the breakdown mill to wire mill 250 will improve the technical and economic indicators of the rolling complex (table 4).

Table 4. Technical and economic indicators of the rolling complex for *Sit*-5.

Indicators	Units	Normative values			
		breakdown mill		wire mill	
		plan	project	plan	project
1. Productive capacity	t/year	6702418	6756418	886627	941930
2. Volume of production	t/year	6620133	6674133	875742	930459
3. Commodity production	t/year	1527584	1581584	871896	926613
4. Mill productivity	t/h	848.3	855.2	119,2	126.7
5. Specific metal consumption	t/t	1.107		1.0277	1.0253
6. Rolling costs	rub/t	73.52	72.82	147.20	139.56
7. Cost of production	rub/t	1768.79	1767.89	1998.3	1984.32
8. Profit of marketable products	mln rub	945.69	978.62	609.82	643.54
9. Product profitability	%	35	36	35	36
10. Annual savings	mln rub	–	6.00	–	13.05
11. Capital investments	mln rub	–	–	–	2.5
12. Project payback period	year	–	–	–	0.2

The developed mathematical models of the cycles of operation of equipment, sections, mills of the rolling complex, the sequence of which is shown in figure 3, is necessary to solve the problems of optimizing many technological routes, as well as for planning and operational management of production.

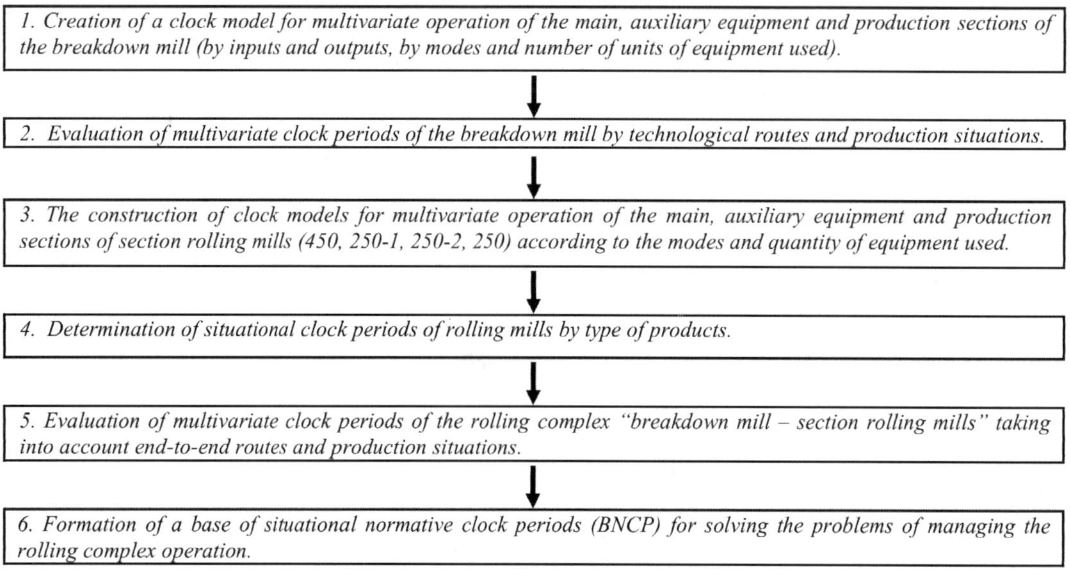

Figure 3. An enlarged diagram of the formation of BNCP.

References

[1] Musatova A I, Kulakov S M and Kadykov V N 2011 *Steel in Translation* vol 41 No 4 294–300
[2] Kulakov S M, Musatova A I et al 2015 *Steel in Translation* vol 45 No 5 338–346

IOP Conf. Series: Materials Science and Engineering **865** (2020) 012006 doi:10.1088/1757-899X/865/1/012006

The dual control algorithm of nonlinear dynamic processes in the conditions of incomplete a priori information

A V Medvedev and A V Raskina

Siberian Federal University, 79 Svobodnyy Ave., Krasnoyarsk, 660041, Russia

E-mail: Raskina.1012@gmail.com

Abstract. The problem of controlling nonlinear dynamic systems under conditions of incomplete a priori information about the model of the differential equation of the process is considered. Nonparametric models are given in which the order of the difference equation of the dynamic process is refined based on the rule for distinguishing significant variables. According to this rule, only those variables are included in the nonparametric model for which the minimum bandwidth is minimal. Nonparametric algorithms of dual control of nonlinear objects are given. Control devices built on the basis of these algorithms perform not only the direct function of controlling the object, but also its study. The process of training the dual control system with active accumulation of information is analyzed. The results of a numerical study of the application of the nonparametric adaptive dual control algorithm are presented.

1. Introduction

Most of the developed control algorithms, including control algorithms for nonlinear dynamic objects, belong to the class of parametric ones [1-3]. This means that at the stage of setting the control problem it is assumed that the parametric structure of the model of the object under study is known, accurate to the parameters from the available a priori information. This approach has received significant development in the framework of the theory of adaptive control systems [3], which involves preliminary parameterization of the object. Adaptive systems in the formation of control actions use current information about the object of study [4, 5]. It should be noted that in most cases the available a priori information is not enough for a reasonable choice of the parametric structure of the model of a dynamic object. Also, one of the characteristic situations is the lack of current information due to the lack of appropriate controls.

Of significant interest is the development of the theory of dual control. The lack of a priori information about the object of study leads to the need to combine the study of the object and control. The control actions are dual in nature. They serve as a means of studying, cognizing an object, but also as a means of bringing the object to the required state. Such control, in which control actions are dual in nature, is called dual control. Dual control was discovered by A.A. Feldbaum in 1960 and developed on the basis of the theory of statistical solutions [6]. It should be noted that learning control systems are not only capable of studying the characteristics of an object, but also, storing them in memory, develop rational control actions. The problem of dual control was also considered by Ya.Z. Tsypkin in [7]. In this statement of the control problem, information is needed on the parametric structure of the model of the object of study. The theory of dual control was further developed in the studies of various authors, in particular B. Wittenmark [8]. In the paper, attention is drawn to the cumbersomeness of analytical results on the synthesis of an optimal control algorithm, and control

IOP Conf. Series: Materials Science and Engineering **865** (2020) 012006 doi:10.1088/1757-899X/865/1/012006

algorithms are replaced in the form of non-optimal dual controllers. Note that in the formulation of the problem, the probability distribution of the interference, as well as the parametric model of the object, are assumed to be known. Such an approach is applicable only to the control of well-studied processes whose parametric structure is known. Some issues related to dual control, namely, the synthesis of dual control systems related to the method of maximum likelihood in the absence of a priori information about unknown parameters of an object, were developed by V.P. Zhivoglyadov in [9]. Later, dual adaptive control algorithms were developed for linear stochastic systems with constant but unknown parameters [10, 11], and various control methods for a linear discrete system with non-stationary random parameters in the presence of interference were studied [12].

In the above studies, control problems were considered under conditions of parametric uncertainty. In the case when the structure of the dynamic process is not defined accurate to the vector of parameters, one of the possible ways to solve the control problem is to use nonparametric adaptive control systems that preserve the components of dualism [13, 14]. This article proposes a nonparametric dual control algorithm, the construction of which is based on the use of a nonparametric regression estimate, which takes into account information about the order of the difference equation of the dynamic process model.

2. Problem set-up

The task of controlling a dynamic object is to form such control actions $u(t)$ that ensure that the output variable of the object $x(t)$ is reduced to the master action $x^*(t)$. In this case, the state of the dynamic system is determined by the values of the controlled input variables $u(t)$, as well as the values of the outputs of the object at previous times.

Let us consider in detail the idea of automatic control of various objects, including dynamic ones, using devices synthesized by the inverse operator method. We introduce the operator of object A, which describes the process

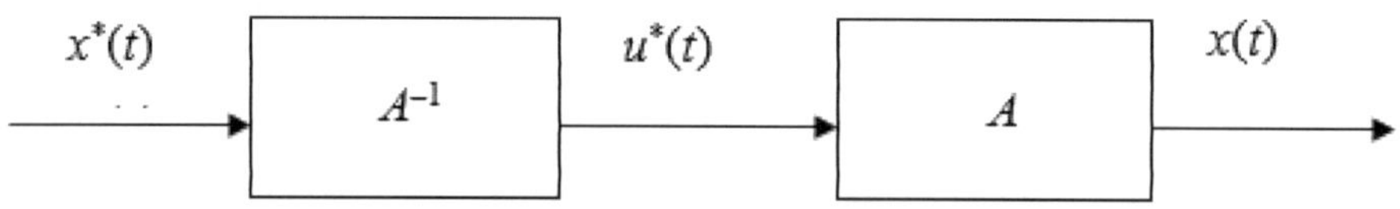

Figure 1. The ideal control scheme.

The idea of control is as follows. This control device converts the vector of measured variables into a vector of control actions, and the operator of such a conversion A^{-1} is the inverse of the operator A, which describes the object:

$$\begin{gathered} x(t) = A < u(t) > \\ A^{-1}x(t) = A^{-1}A < u(t) >, \\ u(t) = A^{-1}x(t). \end{gathered} \quad (1)$$

By setting the desired trajectory of the output variable $x(t) = x^*(t)$, we find from (1) the ideal value of the control action $u^*(t)$. Thus, (1) can be classified as ideal regulators. However, the problem is that in most cases it cannot be built, especially since the operator A is unknown. In the absence of a priori information about the operator A, one of the ways to solve this problem is to use nonparametric control algorithms, which will be shown in detail below.

Figure 2 shows the control block diagram of the process. In figure 2, the following notation is used: (t) is continuous time, index t is discrete time, random interference measurements h_t^u, h_t^x of the process variables, $\xi(t)$ is a vector random noise. Variables are monitored over a time interval Δt.

The paper considers classes of control objects that can be described by difference equations of the form:

$$x_t = F(x_{t-1}, \ \ldots, \ x_{t-k}, \ u_t, \xi_t). \quad (2)$$

where F is an unknown functional, k is the order of the difference equation, which is bounded by $\leq k_{\max}$. The input and output of a dynamic object is represented by measurements forming a sample of the form $\{u_i, x_i\}, i = \overline{1, s}$, where s is the sample size, u_i, x_i- measures the input and output of the object at time t_i.

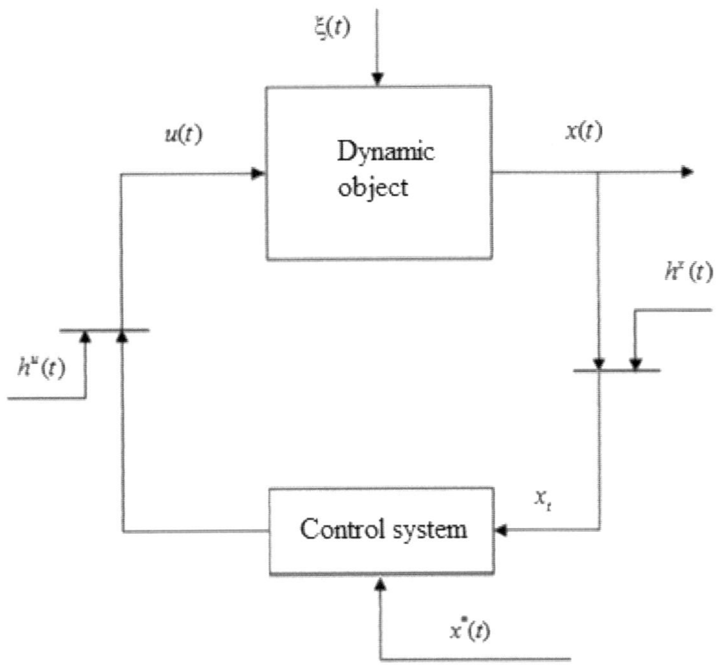

Figure 2. Dynamic object control scheme.

The following nonparametric estimate of the regression function based on observational data $\{x_i, u_i, i = \overline{1, s}\}$can be accepted as a nonparametric model of the object [13]:

$$x_s^t = \frac{\sum_{i=1}^{s} x_i \cdot \Phi\left(\frac{u_s - u_i}{c_s^u}\right) \prod_{j=1}^{k} \Phi\left(\frac{x_{s-j} - x_{i-j}}{c_s^{x[j]}}\right)}{\sum_{i=1}^{s} \Phi\left(\frac{u_s - u_i}{c_s^u}\right) \prod_{j=1}^{k} \Phi\left(\frac{x_{s-j} - x_{i-j}}{c_s^{x[j]}}\right)}, \quad (3)$$

where $\Phi(\cdot)$ is a kernel function, $c_s^u, c_s^{x[j]}$- is a bandwidth. The optimal bandwidths are found by minimizing a quadratic error function (4) by using the sliding exam method

$$R(c_s^u, c_s^{x[1]}, \ldots, c_s^{x[k]}) = \sum_{q=1}^{s} \left(x_s(u_q, x_{q-1}, \ldots, x_{q-k}) - x_q\right)^2 = \min_{c_s^u, c_s^{x[1]}, \ldots, c_s^{x[k]}}, q \neq i, \quad (4)$$

where i is the index in the equation (3).

3. Definition of significant variables

It is significant in the estimate (3) that, in accordance with each output delayed variable $x_{s-1},\ldots,x_{s-k}$, some values of the bandwidth $c^{*x^1}_s,\ldots,c^{*x^2}_s$. Let us give an example, let $c^{*x^1}_s$ be the bandwidth for x_{s-1}, and $c^{*x^2}_s$ for x_{s-2}. If $c^{*x^1}_s < c^{*x^2}_s$ then $\Phi\left(\frac{x_{s-1}-x_{t-1}}{c^{*x^1}_s}\right) < \Phi\left(\frac{x_{s-2}-x_{t-k}}{c^{*x^2}_s}\right)$, and then the output variable x_{s-1} has a greater effect on the output quantity than x_{s-2}. Graphically, this is expressed as follows (figure 3).

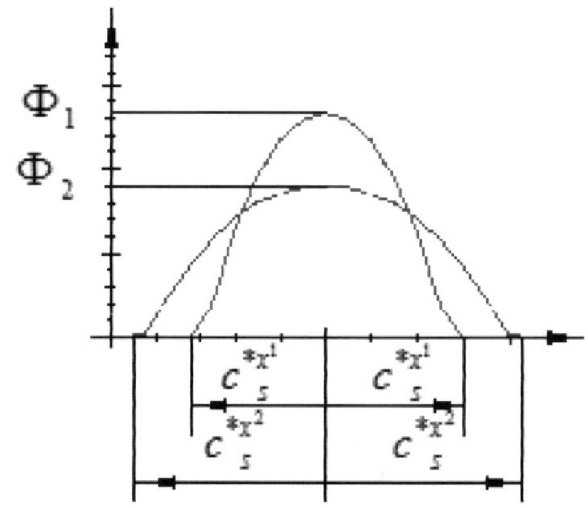

Figure 3. The dependence of the value of nuclear function $\Phi(\cdot)$of the bandwidths c_s.

In the figure 3: $\Phi_1 = \Phi\left(\frac{x_{s-1}-x_{t-1}}{c^{*x^1}_s}\right), \Phi_2 = \Phi\left(\frac{x_{s-2}-x_{t-k}}{c^{*x^2}_s}\right)$.

The algorithm for calculating significant variables x_{s-j} is based on the following scheme. First, the initial value of k is given. The model is constructed by equation (4) and the relative error W_0 is calculated:

$$W = \sqrt{\frac{1}{s}\sum_{i=1}^{s}(x_i - x_i^s)^2 \Big/ \sum_{i=1}^{s}\frac{1}{s-1}(m_x - x_i)^2}$$

where m_x is an expected value.

For each i - th iteration, the following set of actions is performed:

1. For each coefficient the optimal value is found: $c_s^{x^1} = c^{*x^1}_s, c_s^{x^2} = c^{*x^2}_s, \ldots, c_s^{x^k} = c^{*x^k}_s$
2. The maximum of all the values obtained is found: $c^{x^j}_{\max\_s}$
3. The model is constructed by the equation (4). The multiplier $\mathrm{H}\left(\frac{x_{s-j}-x_{t-j}}{c^{*x^j}_s}\right)$is excluded, taking into account that j is a number for $c^{x^j}_{\max\_s}$
4. A relative error W_i is calculated.

These actions will be repeated until $W_i \geq W_{i-1}$.

4. Nonparametric dual control algorithm

The nonparametric dual control algorithm has the form [13,14]:

$$u_{s+1} = u_s^* + \Delta u_{s+1}, \tag{5}$$

where u_s^* is the component that accumulates information about the object of study, and $\Delta u_{s+1} = \varepsilon(x_{s+1}^* - x_s)$ are the learning search steps.

The dual control scheme is shown in figure 4.

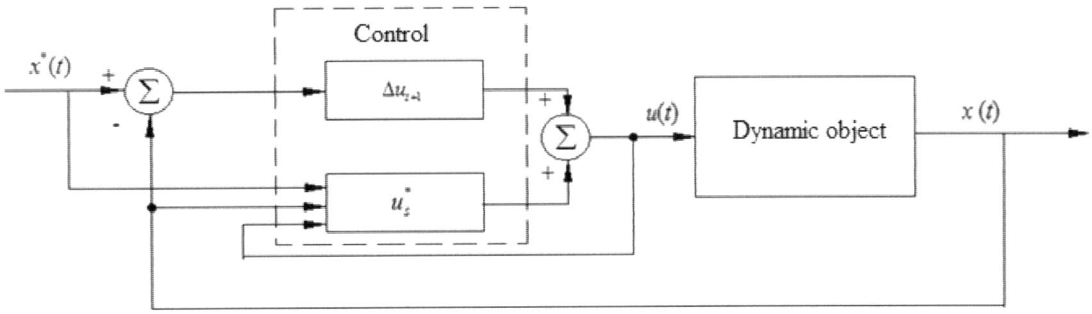

Figure 4. Diagram of dual control of a dynamic object.

The dualism of algorithm (5) is as follows. At the first control steps, the main role in the formation of control actions is played by the term Δu_{s+1} from formula (5). But already with the accumulation of information about the object, the role of the term u_s^* increases.

In this case, as the term u_s^* from equation (5), we can take the expression (6)

$$u_s^* = \frac{\sum_{i=1}^{s} u_i \cdot \Phi\left(\frac{x_{s+1}^* - x_i}{c_s}\right) \prod_{j=1}^{k} \Phi\left(\frac{x_{s-j} - x_{i-j}}{c_s}\right)}{\sum_{i=1}^{s} \Phi\left(\frac{x_{s+1}^* - x_i}{c_s}\right) \prod_{j=1}^{k} \Phi\left(\frac{x_{s-j} - x_{i-j}}{c_s}\right)}. \quad (6)$$

The control algorithm of nonlinear dynamic systems is constructed as follows. Based on the rule for distinguishing essential variables, the order of the spread equation of the dynamic process model k is determined, which is further used in the calculation of control actions in (5), where only those variables that were selected by the algorithm are present.

In this experiment, the quality of control was evaluated according to two characteristics:

- The regulation time t_p is the time from the start of control until the moment when the output value differs from the task no more than some given value α. Usually taken α= 0.05*x*.
- The relative control error W_p is equal to the total deviation of the actual process output from the setpoint during the entire control time with respect to the setpoint, expressed in relative values in %.

$$W_P = \frac{\frac{1}{s}\sum_{i=1}^{s} |x_i - x_i^*|}{x^*}$$

Shows the degree to which the output value deviates from the reference, expressed as a percentage.

Figure 5 shows the results of a computational experiment for controlling a nonlinear dynamic system $x_t = 0{,}2x_{t-1}^{1.2} - 0{,}3x_{t-2}^{1.2} + 0{,}4\cos(x_{t-3}) + 1{,}5u_t$

The relative control error for the case shown in Figure 5 is W_p=0.216. Figure 6 shows an experiment on controlling a dynamic object described by the difference equation $x_t = 0{,}2x_{t-1}^{1.2} - 0.3x_{t-2} + 0{,}4x_{t-3} + 1{,}5u_t$ under external noise $\xi(t)$ =5 %.

In Fig. 6, the relative control error is W_p=0.162.

The developed nonparametric algorithm for controlling a nonlinear dynamic system allows you to effectively control processes, even in the presence of external noise.

IOP Conf. Series: Materials Science and Engineering **865** (2020) 012006 doi:10.1088/1757-899X/865/1/012006

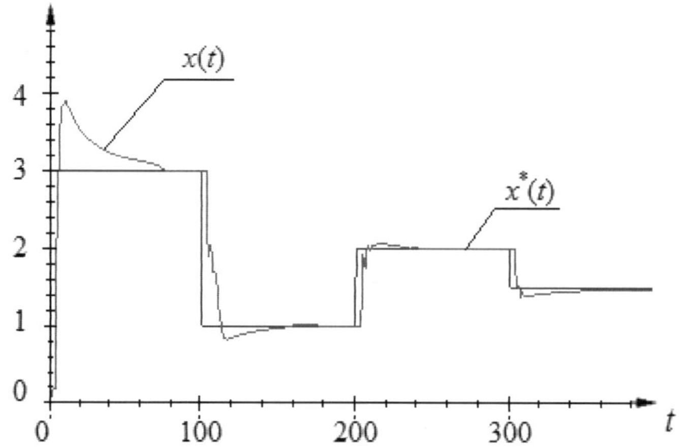

Figure 5. Control result with unknown parametric structure of a nonlinear dynamic object.

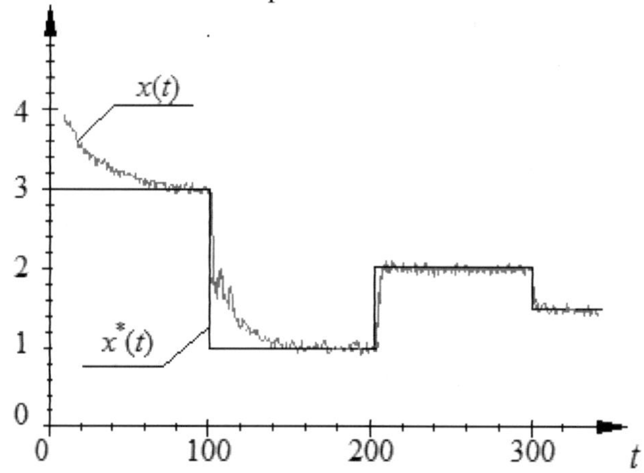

Figure 6. Control result with unknown parametric structure of a nonlinear dynamic object, at a noise $\xi(t)$ =5 %.

5. Conclusion

The article provides a solution to the control problem for the case of nonlinear dynamic systems. Nonparametric algorithms, as was previously known, can be effectively applied to solve identification and control problems for processes from a class or category of linear ones. Naturally of interest is the situation when the process under investigation is non-linear, to one degree or another. In the article, a nonlinear dynamic system was considered as a system at the input of which output variables were delayed by the corresponding number of clock cycles. Moreover, the model of a dynamic system was a nonparametric estimate of the regression function from observations. To improve this model, an algorithm for determining the order of the difference equation of the dynamic process model was proposed. The algorithm reduces to determining the delayed components of the output variable of the model of the difference equation of a dynamic object. One of the main advantages of the proposed method in comparison with the currently dominant approaches is the fact that the developed nonparametric algorithm is more applicable to practical tasks, as it is able to work in conditions of small a priori information about the object.

The computational experiments presented above showed that a nonparametric control algorithm can also be used for objects for which the principle of superposition is no longer fulfilled.

Reference

[1] Pupkov K A and Egupov N D 2004 *Methods of the Classical and Modern Theory of Automatic Control. T1: Mathematical Models, Dynamic Characteristics and Analysis of Control Systems* (Moscow: M57GTU n.a. N.E. Bauman), p 656 (In Russia)

[2] Dorf R and Bishop R 2002 *Modern Management Systems* (Moscow: Laboratory of Basic Knowledge) p 832

[3] Tsykunov A 2018 *Adaptive and Robust Control of Dynamic Objects by Exit* (Moscow: Litres)

[4] Song X, Ahmadian M and Southward S 2003 *An Adaptive Semiactive Control Algorithm for Vehicle Suspension Systems*. In ASME 2003 International Mechanical Engineering Congress and Exposition. American Society of Mechanical Engineers Digital Collection pp 219–228

[5] Fei J and Ding H 2012 *Nonlinear Dynamics* **70(2)** 1563–73

[6] Feldbaum A A 1966 *Fundamentals of the Theory of Optimal Automatic Systems* (Moscow: Nauka)

[7] Tsypkin Ya Z 1968 *Adaptation, Training and Self-training in Automatic Systems* (Moscow: Nauka)

[8] Wittenmark B 1995 *IFAC Adaptive Systems in Control and Signal Pricessing* **28(13)** 67–72

[9] Zhivoglyadov V P and Medvedev A V 1974 *Nonparametric Adaptation Algorithms* (Frunze: Ilim)

[10] Li D, Qian F and Fu P 2008 *Automatica* **44(1)** 119–127

[11] Wenk C and Bar-Shalom Y 1980 *IEEE Transactions on Automatic Control* **25(4)** 703–710

[12] Tse E and Bar-Shalom Y 1973 *IEEE Transactions on Automatic Control* **18(2)** 109–117

[13] Medvedev A V 2015 *Fundamentals of the Theory of Adaptive Systems* (Krasnoyarsk: SibSAU) p 525

[14] Medvedev A V 2018 *Fundamentals of the Theory of Nonparametric Systems. Identification, Management, Decision Making* (Krasnoyarsk: SibSAU) p 729

Automated test bench for researching the characteristics of an AC motor based on the altivar variable-frequency drive

V A Kubarev, T V Bogdanovskaya, O A Ignatenko and O R Gallyamova

Siberian State Industrial University, 42 Kirova Street, Novokuznetsk, 654006, Russia

E-mail kubarev.sibsiu@list.ru

Abstract. The article is devoted to the development of an automated test bench for studying the characteristics of an alternating current motor based on the Altivar variable-frequency drive, which allows automatic or manual recording of various characteristics of an asynchronous motor. A block diagram of the test bench and its technical structure are presented, algorithms for the functioning of the software are developed.

1. Introduction

Currently, due to increased requirements for production efficiency, industrial enterprises tend to replace adjustable DC electric drives with adjustable drives based on squirrel-cage induction motors. Moreover, in the near future, with the inevitable increase in energy prices, the number of installed frequency converters, and, accordingly, asynchronous motors will only grow from year to year [1, 2].

Asynchronous drive is actively conquering the market, so now it is extremely important to educate students and improve the skills of engineering personnel. In addition, the current Federal state educational standards of both higher and secondary vocational education are aimed at the fact that the training of specialists of secondary vocational education, as well as bachelors, should for the most part be aimed at acquiring practical skills, that is, conducting practical and laboratory work, which requires modern laboratory facilities [3, 4].

A modern laboratory test bench should allow the study of an asynchronous squirrel-cage electric motor, in particular, the following characteristics should be investigated:

- natural mechanical and electromechanical characteristics.
- energy diagrams.

Carry out an open-loop study of the VDF - asynchronous motor (VFD – AM) and take:

- mechanical characteristics;
- adjusting characteristics;
- energy diagrams;

Carry out a study of the closed-loop system VFD – AM:

- with setting the speed control loop;
- with the study of a closed system;
- take energy diagrams.

In addition, the following additional signaling devices must be present at the test bench:

- switching on the QF1 — test bench ON;
- switching on the contactor KM1 — VFD ON;
- switching on the ET6 load converter;

- emergency stop of the operation (shutdown of ET6 DC drive or VFD).

2. Results and discussion

For these purposes and taking into account previously defined requirements, a stand was designed at the Department of Electrical Engineering, Electric Drive and Industrial Electronics to replace the existing laboratory installation of a frequency-controlled electric drive VFD-AM in the discipline "Electric Drive" for laboratory work on the discipline "Electric Drive" in the areas of training "Electric Drive and Automation" and "Electronics and Nanoelectronics". The proposed laboratory test bench allows you to manually or automatically record the characteristics of an induction motor.

Figure 1 shows a block diagram of an automated laboratory test bench.

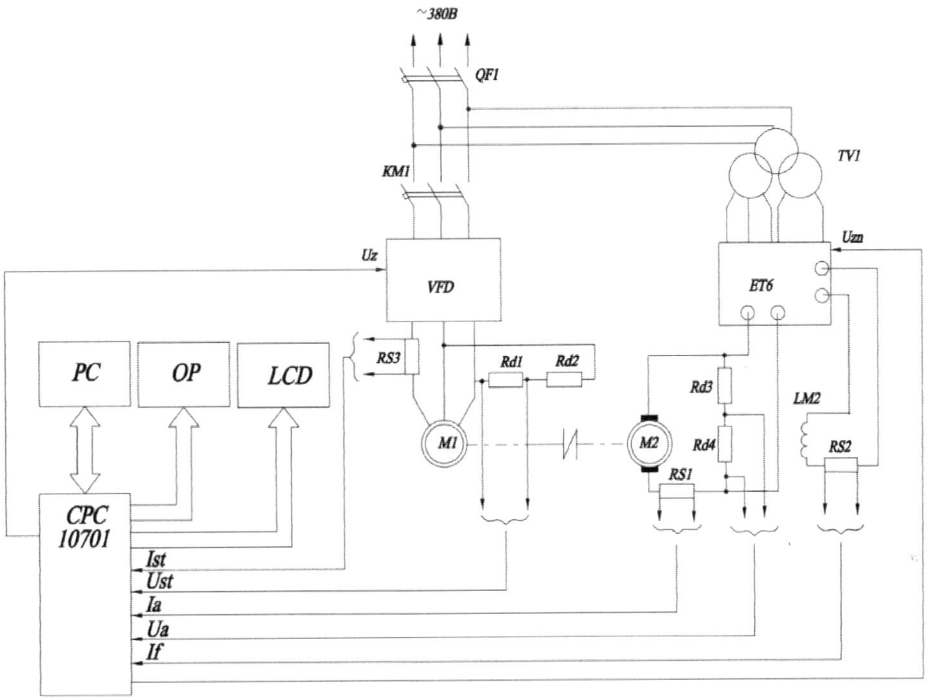

Figure 1. Block diagram of an automated laboratory test bench.

The test bench includes:

- Programmable Fastwel CPC10701 controller with a control panel (OP), an LCD display (LCD) and a power supply (PW);
- Asynchronous AC motor (M1);
- DC load motor (M2);
- Variable-frequency drive Altivar 5 (VFD) [5];
- Thyristor converter ET-6 (ET6) [6];
- The power part, including an automatic switch (Q1), a control contactor (KM1), a transformer of a DC drive (TV1), as well as current sensors (RS1, RS2, RS3) and voltage in the circuits of the tested and load motors.
- Personal computer — a student's workplace (PC) with software installed on it.

Based on the presented functional diagram, the technical structure of the stand hardware was developed, with the possibility of conducting an automated experiment, shown in figure 2. This structure shows the technical connections between the devices included in the data acquisition and control system and external systems.

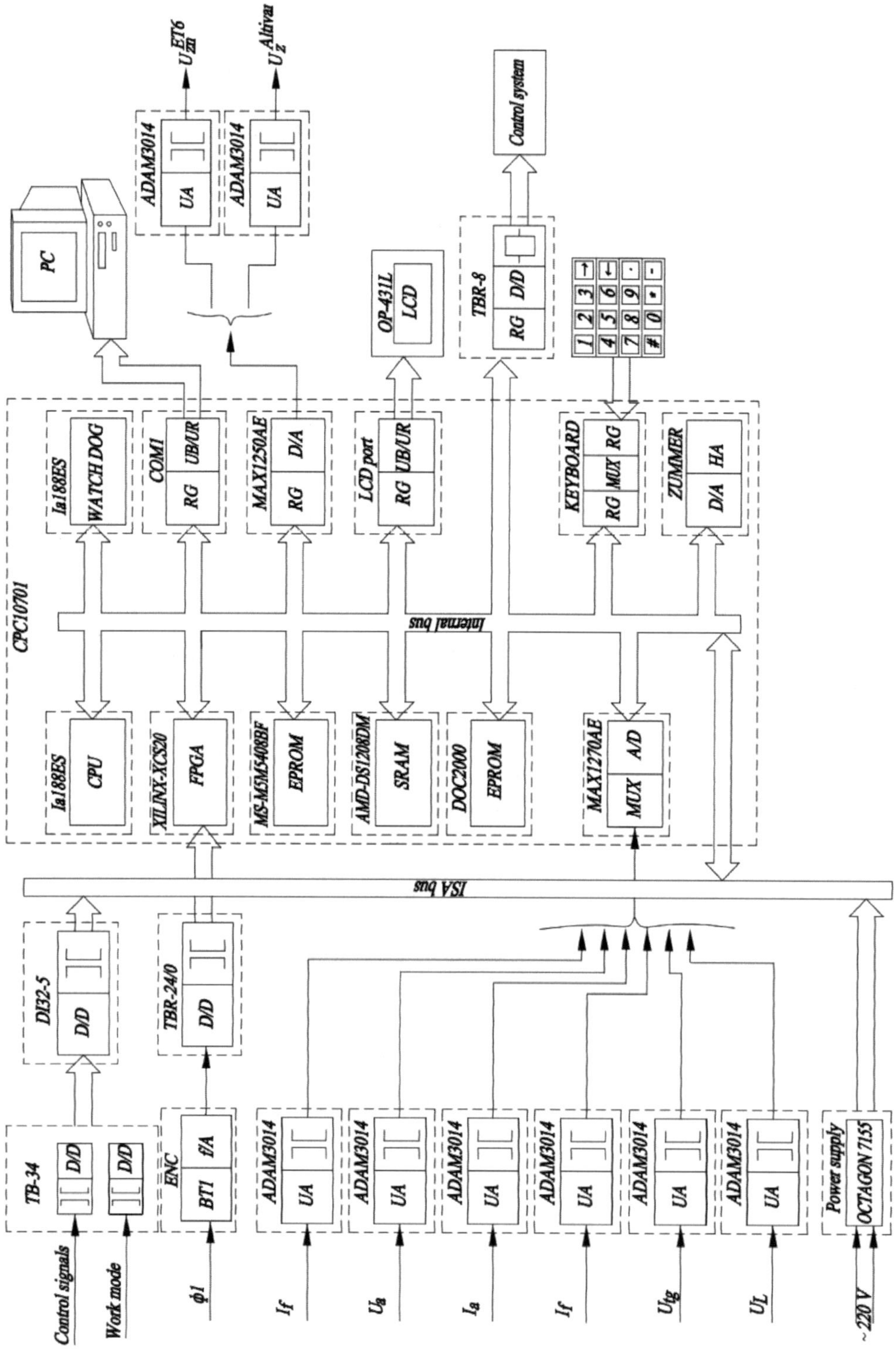

Figure 2. Technical structure of the laboratory test bench.

Software was developed for an automated data collection system of the stand. The software was created on the principle of modularity, which implies that each independent function is a separate module that has its own input and output parameters and operates independently of other modules based only on the values of the input parameters [7]

To obtain the normalized execution time of the program cycle, and guaranteed metrological characteristics of the system, the program is divided into two parts: the called and the calling. The calling part works with the data collection equipment, and all the basic functions of the system that require metrological regulation are met, and the functions of working with the user and information processing are located in the called part. The algorithm of the main program with a timer binding is shown in figure 3, and the algorithm of the subroutine for working with signals and generating control actions is shown in figure 4.

On the basis of the presented stand, it is planned to conduct laboratory work on the following topics:

- Experiment "Study of the mechanical and electromechanical characteristics of an asynchronous motor with a phase rotor" – a study of the main characteristics of an asynchronous motor.
- Experiment "Study of the main characteristics of a frequency-controlled asynchronous electric drive" – a study of the main characteristics of an asynchronous motor with frequency regulation.

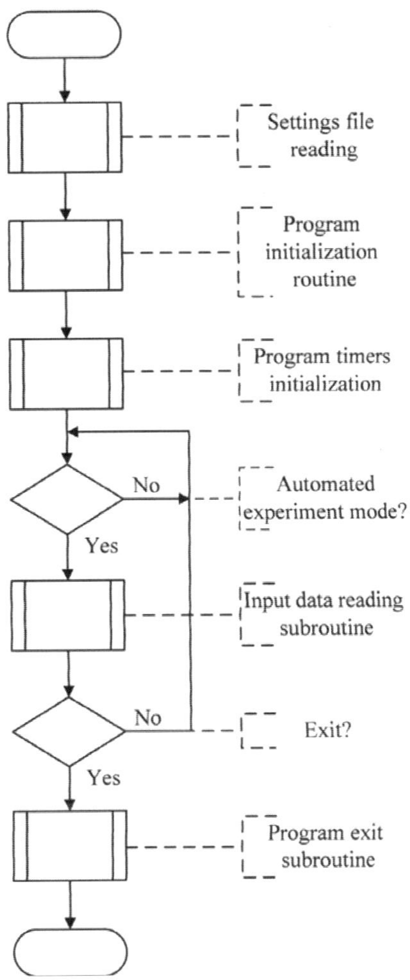

Figure 3. Timer-bound main program algorithm.

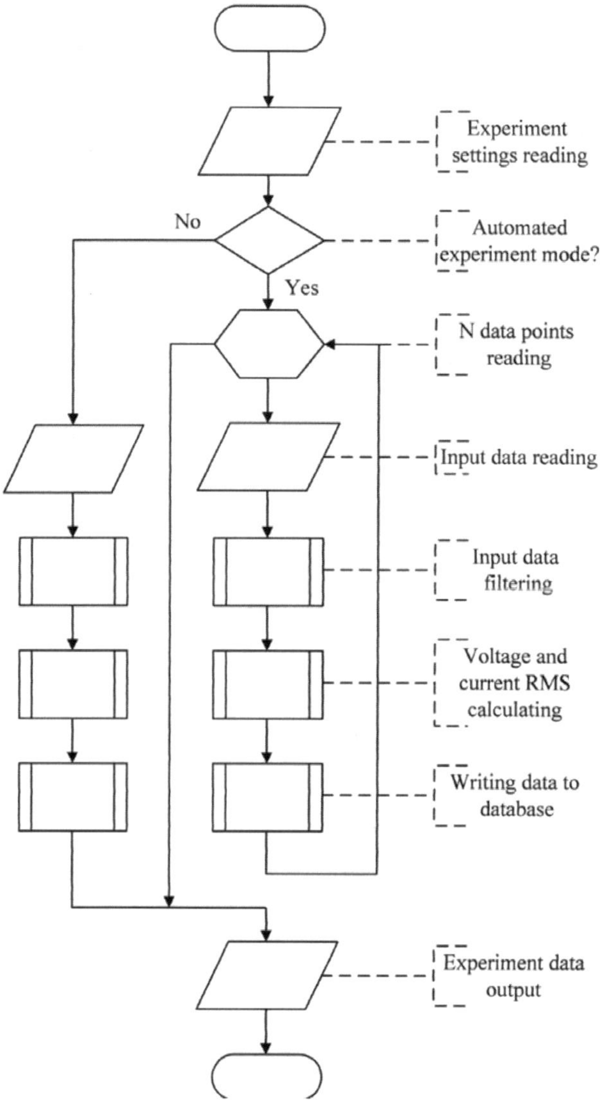

Figure 4. The algorithm of the subroutine to work with signals, and generate control actions.

3. Conclusion

A laboratory test bench is proposed that allows, in an automated or manual mode, to take various characteristics of asynchronous motors. The proposed approach to the construction of a laboratory bench and its electric drive control system allows you to speed up the process of developing such systems.

References

[1] Anuchin A S 2019 *Electric Drive Control Systems* (Moscow: MEI Publishing House)

[2] Fashchilenko V N 2011 *Adjustable Electric Drive of Pumps and Ventilators of Mining Enterprises* (Moscow: Mountain Book)

[3] Federal State Educational Standard of Higher Education in the Field of Training 13.03.02 Electricity and Electrical Engineering (Approved by Order of the Ministry of Education and Science of Russia No. 955 of September 3, 2015)

[4] Federal State Educational Standard of Higher Education in the Field of Training 13.04.02

Electricity and Electrical Engineering (Approved by Order of the Ministry of Education and Science of Russia No. 1500 of November 21, 2014)
[5] *Altivar 5 series 45. Variable speed controllers for asynchronous motors. User's manual* (Groupe Schneider 1993)
[6] *Direct current electrodrive ET6. User's manual.*
[7] Ostrovlyanchik V Yu and Kubarev V A 2012 *Science Bulletin of the NSTU* **4** 186–190

IOP Conf. Series: Materials Science and Engineering **865** (2020) 012008 doi:10.1088/1757-899X/865/1/012008

Operational algorithm of logical control system for electric drive of a lifting installation based on a doubly-fedinduction machine

V Yu Ostrovlyanchik and I Yu Popolzin

Siberian State Industrial University, 42 Kirova str., Novokuznetsk, 654 007, Russia

E-mail: eidoline@yandex.ru

Abstract. Using of a doubly-fed machine circuit is a promising way to upgrade existing electric drives of hoist installations based on wound-rotor induction motors.Power connection of induction motors according to the scheme of a doubly-fed machine with control of machine coordinates in the rotor circuit allows to increase the energy efficiency of the drive, organize the recovery of energy, which is usually uselessly dissipated in the rotor circuit, as well as increase the reliability of the drive. To achieve a wide range of speed change while maintaining the overload capacity of the machine by the torque, it is necessary to combine various methods for controlling the amplitude, phase and frequency of the auxiliary voltage on the rotor.These parameters are controlled by a logical control system. So it is necessary to develop an operational algorithm of the logical control system to organize this way of control. This paper is devoted to the development of the operational algorithm of the logical control system for the electric drive of a lifting installation based on a doubly-fedinduction machine, which allows combining different modes of operation of the machine to achieve a wide range of speed control.

1. Introduction

Doubly-fed machine (DFM) in the general case is a wound-rotor induction motor (WRIM) with auxiliary voltage supply in a rotor circuit from an external source with the options to control the amplitude, frequency and phase of additional voltage [1].

The use of DFM in adjustable-speed drives has significant advantages over the use of WRIM with auxiliary resistors in the rotor circuit, since it allows controlling the flow of slip energy and achieving high energy efficiency. Nevertheless DFM operation in electric drive systems requires switching of its operation modes depending on the rotor speed. The switching of modes is carried out by the logical control system of the electric drive.

Since the modernization of complex technological complexes should not affect their operability [2], the use of a DFM scheme is the best way to modernize existing lifting systems based on WRIM.

In [3] it was shown that in order to achieve a wide range of DFM speed change it is necessary to combine its operation modes. A general algorithm of the drive functioning for this case was obtained in [4].

2. Electric drive structure

The algorithmic structure of the electric drive of a lifting installation with DFM was obtainedin [6] and [6]. The structure shown in figure 1.

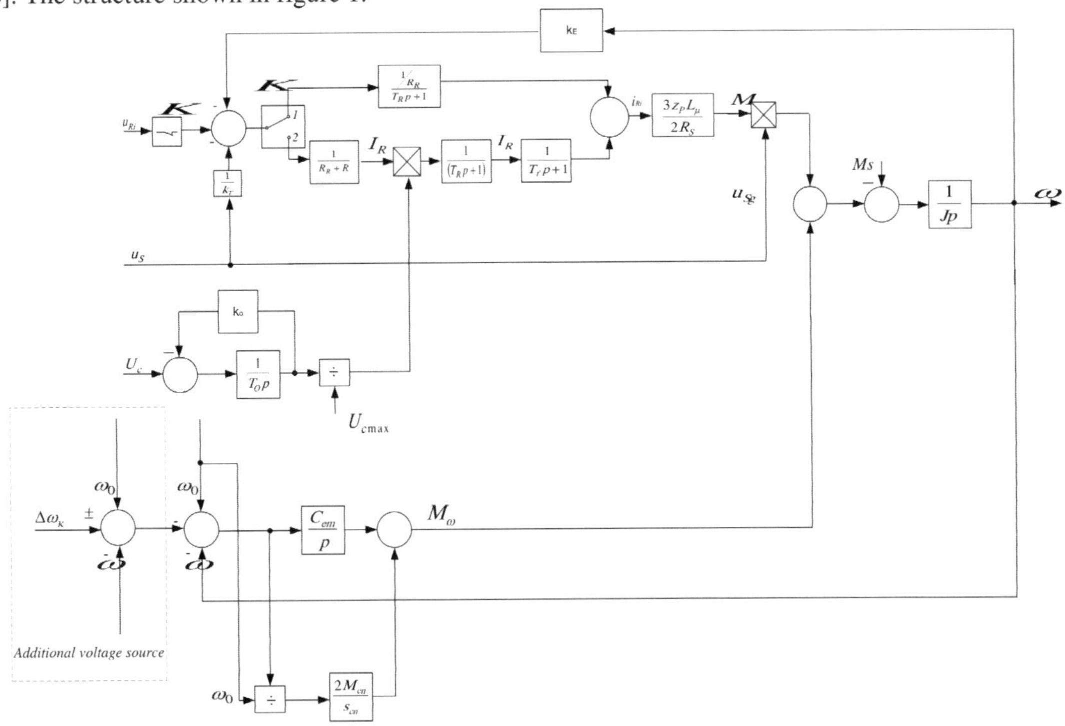

Figure 1. The algorithmic structure of the electric drive of a lifting installation with DFM.

Notation for figure 1: u_{Ri}- active component of the additional voltage on the rotor; R_R – rotor intrinsic resistance; T_R - electromagnetic time constant of rotor circuit; k_E – rotor EMF internal feedback coefficient; i_{Ri} - active component of rotor current; z_P – the number of pole pairs of the motor; L_μ – mutual inductance between the stator and rotor; R_S – stator intrinsic resistance; u_{Sg}–stator voltage (phase-to-neutral); U_C–control voltage of the additional active resistance in the rotor circuit; C_{em} – electromagnetic stiffness; M_{cn} – critical torque on the WRIM natural mechanical characteristic; s_{cn} – critical slip on the WRIM natural mechanical characteristic; k_T – WRIM transformation ratio; M – electromagnetic torque; M_c – load torque (static torque); J – the total moment of inertia of the mechanical part of the installation reduced to the rotor axis; ω – rotor angular velocity; ω_0 – synchronous angular velocity; $\Delta\omega_K$ – additional voltage frequency increment; K – a logical function that determines whether the drive operates in doubly-fed mode from a controlled voltage converter (CVC) or in a controlled current converter (CCC) mode [7].

Let us assume that position 1 of key K corresponds to logical 0, and 2 corresponds to logical 1. In this case, the DFM operation modes are determined by the following expressions:

$$\begin{cases} K = 0 - CVC; \\ K = 1 - CCC. \end{cases} \quad (1)$$

We introduce the following notation: $\omega_н$ – rated speed of the DFM on a natural mechanical characteristic; ω_t – DFM speed set by the control system; ω_{cr}=$0{,}5\omega_0$ – speed corresponding to the switching from asynchronous to synchronous mode; A – maximum rate of change in the amplitude of the additional voltage on the rotor, V/s; U_R– effective value of the additional voltage on the rotor. It is possible to draw up rules for controlling the additional voltage on the rotor of the machine,based on the analysis of the mechanical characteristics of the machine in various modes. The control rules assume that the working area of the DFM mechanical characteristics is divided into sections within the mechanical characteristics of the machine do not change in a qualitative way.

A special mode of operation of the machine is brake release, in which the drive provides a torque when the rotor is locked. This mode occurs when a mechanical brake is applied to the drum of a lifting machine, what is mathematically equivalent to the condition $J=\infty$. This mode is described by the following system of equations:

$$\begin{cases} e_{SgRi,} = u_S \dfrac{\cos\varphi_S}{k_T}, \\ \Delta u_{Ri} = u_{Ri} + e_{SgRi}, \\ i_{Ri} = \Delta u_{Ri} \cdot \dfrac{1/R_R}{T_R p + 1}, \\ M_a = \dfrac{3 z_P L_\mu}{2 R_S} i_{Ri} u_{Sg}. \end{cases} \quad (2)$$

System of equations (2) can be associated with the structural scheme of a DFM with a locked rotor, shown in figure 2.

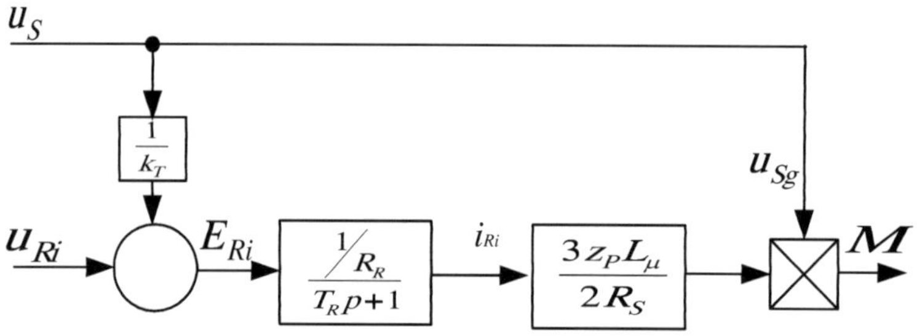

Figure 2. Structural scheme of a DFM with a locked rotor.

In the structures shown in Figures 1 and 2, two parts can be distinguished: variable one, which determines the dependence of the electromagnetic torque of the machine on the parameters of the additional voltage on the rotor or its active resistance $M=f(U_R,\delta,R_R)$, that is, characterizing the electromechanical conversion of energy in a machine; and non-variable one, determining the dependence $\omega=f(M)$, i.e. mechanical energy conversion. Thus, the main idea of controlling a combined drive is to change the method of electromechanical energy conversion in a machine while maintaining the mechanical conversion method.

3. Functional algorithm

In accordance with the known sections of the working cycle of the lifting installation, we can formulate the rules for switching inside and between them and compose analytical and logical expressions describing the algorithm of the logical control system.

We additionally introduce the following variables for the description of proposed algorithm:

I_{Ra}– active component of rotor current;

$f(K)$ – rotor circuit key closure variable;

S – a variable that takes a value of 1 in the presence of a drive start signal from a stopped state and a value of 0 in the absence of this signal;

B – a variable that takes a value of 1 in the presence of a brake signal of the drive and a value of 0 in the absence of this signal;

St – a variable that takes a value of 1 in the presence of a drive stop signal and a value of 0 in the absence of this signal;

Rev – a variable that takes a value of 1 in the presence of a drive reverse signal and a value of 0 in the absence of this signal;

CR – variable taking value 1 when switching the reverser;

$ES=\begin{cases}1 \; when \; \omega=\omega_t, \\ 0 \; when \; \omega\neq\omega_t\end{cases}$ – variable determining the machine reaches a given rotor speed;

$CD=\begin{cases}1 \; when \dfrac{dI_{Ra}}{dt}=0, \\ 0 \; when \dfrac{dI_{Ra}}{dt}\neq 0.\end{cases}$ – variable determining the equality of 0 time derivative of the active component of the rotor current;

$CS=\begin{cases}1 \; when \; I_{Ra}>0, \\ 0 \; when \; I_{Ra}\leq 0.\end{cases}$ – variable determining the sign of the active component of the rotor current;

$SC=\begin{cases}1 \; when \; \omega\leq\omega_{KP}, \\ 0 \; when \; \omega>\omega_{KP}.\end{cases}$ – variable determining the machine reaches a critical rotor speed$

$ZS=\begin{cases}1 \; when \; \omega=0, \\ 0 \; when \; \omega\neq 0.\end{cases}$ – variable determining the machine reaches zero rotor speed.

Taking into account the introduced notation, the functioning algorithm of the combined electric drive of the lifting machine based on DFM can be formulated as follows:

1) Brake release and acceleration:

a) Voltage supply connection:

$$\begin{cases}\overline{E}_S=\overline{E}_R+\overline{U}_R, \\ \delta=\psi_{E_R}+180°, \\ I'_R=0, \\ f(K)=1.\end{cases} \tag{3}$$

b) Brake release:

$$At \;\; (\omega=0)\wedge(M=M_C):\begin{cases}K=1, \\ S=1 \\ \dfrac{dI'_R}{dt}>0, \\ M>M_c.\end{cases} \tag{4}$$

c) Acceleration:

$$\begin{gathered} At \;\; 0<\omega\leq 0{,}1\omega_0:\begin{cases}K=1 \\ I'_R=const.\end{cases} \\ At \;\; 0{,}1\omega_0<\omega\leq\omega_0:\begin{cases}K=0, \\ \Delta\omega=\dfrac{M_H s_{KPe}\omega_0-0{,}5(\omega_0-\omega)M_{KPe}}{2M_{KPe}}, \\ -A\leq\dfrac{dU_R}{dt}<0, \\ \delta=arctg\left(\dfrac{s}{s_{KPe}}\right).\end{cases}\end{gathered} \tag{5}$$

Section end condition: $ES \wedge CD \wedge CS = 1$.

2) Movement at low speed:

$$At \quad 0 < \omega \leq 0{,}1\omega_0 : \begin{cases} K = 1 \\ I'_R = const. \end{cases}$$

$$At \quad 0{,}1\omega_0 < \omega \leq 0{,}5\omega_0 : \begin{cases} K = 0, \\ \Delta\omega = \dfrac{M_н s_{KPe}\omega_0 - 0{,}5(\omega_0 - \omega)M_{KPe}}{2M_{KPe}}, \\ U'_R = const, \\ \delta = arctg\left(\dfrac{s}{s_{KPe}}\right). \end{cases} \quad (6)$$

Section end condition: $ES = 0$.

3) Movement at maximum speed:

$$At \quad \omega = \omega_3 : \begin{cases} K = 0, \\ \overline{U}_R = 0. \end{cases} \quad (7)$$

Section end condition: $B=1$.

4) Slowdown:

$$At \quad 0 \leq \omega \leq \omega_0 : \begin{cases} K = 0, \\ 0 < \dfrac{dU_R}{dt} \leq A, \\ f_R = \dfrac{\omega_0 - \omega}{\omega_0} f, \\ \delta = arctg\left(\dfrac{s}{s_{KPe}}\right). \end{cases} \quad (8)$$

Section end condition: $ES \wedge CD = 1$;

5) Movement at low speed:

$$At \quad 0 < \omega \leq 0{,}1\omega_0 : \begin{cases} K = 1 \\ I'_R = const. \end{cases}. \quad (9)$$

Section end condition: $St=1$;

6) Stopping:

$$\begin{cases} K = 1 \\ \dfrac{dI'_R}{dt} < 0. \end{cases}. \quad (10)$$

Section end condition: $ZS \wedge CS = 1$;

7) Reverse: reversing switch when

$$\begin{cases} \overline{E}_S = \overline{E}_R + \overline{U}_R, \\ \delta = \psi_{E_R} + 180°,. \\ I'_R = 0, \\ f(K) = 1. \end{cases} \quad (11)$$

The proposed algorithm (3-11) can be transformed into a block diagram or transition graph.

4. Conclusion

The algorithm involves the regulation of the active component of the rotor current as the basis for regulating the motor torque and maximizing it. To increase the drive efficiency and reduce the total power consumed by DFM, it is also possible to regulate the reactive component of the rotor current or stator current. The logical control system (LCS) implements this algorithm together with a programmable task and control apparatus, which performs the calculation of logical functions based on signals from the electric drive system and generates control actions for the LCS.

References

[1] Onishchenko G B and Lokteva I L 1979 *Asynchronous Valve Stages and Double-fed Motors* (Moscow: Energiya) p 200
[2] Myshlyaev L P et al *2018 IOP Conference Series: MSE* **354** 012008
[3] Ostrovlyanchik V Yu et al *2018 IOP Conference Series: EES* **206** 142714
[4] Ostrovlyanchik V Yu et al 2018 *IOP Conference Series: EES* **206** 012040
[5] Ostrovlyanchik V Yu and Popolzin I Yu 2019 *IOP Conference Series: EES* **377** 012042
[6] Ostrovlyanchik V Yu and Popolzin I Yu 2019 *IOP Conference Series: EES* **377** 012041
[7] Ostrovlyanchik V Yu et al 2019 *Mining Equipment and Electromechanics* **1** 40–46

IOP Conf. Series: Materials Science and Engineering **865** (2020) 012009 doi:10.1088/1757-899X/865/1/012009

Improvement of the regulation system for the electric drive of rolls rotation in the rolling mill

V A Kuznetsov, E S Kuznetsova, V N Kharenko and N C Zaitsev

Siberian State Industrial University, 42 Kirova Street, Novokuznetsk, 654007, Russia

E-mail: Vladimir.kuznezoff@yandex.ru

Abstract. An improved dual-zone speed control system for the electric rotation of the rolls of a metallurgical plant using a dynamic current sensor has been developed. For example, the rolling engine of the mill "900" of the beam workshop was used. The use of an improved system will increase the productivity of the rolling mill.

1. Introduction

Present day in rolling production of metallurgical plants, widespread use has been found for automatic electric drives for rotating rolls (main drives) using the Controllable AC-DC converters using tryristors and typical systems of subordinate control [1]. The rolling engine has the following nominal parameters: frequency – 50 hz, power – 7100 kW, armature voltage – 900 V, armature current – 8100 A, speed – 80/120 rpm, exciting current – 360 A, overcurrent ratio (shutdown) – 2.75.

2. Development of an improved dual-zone speed control scheme

In a typical scheme, to change the speed to the main one, the armature current control loops with an IP-controller and speed with a P-controller are used. To control the speed of the electric motor above the main one, the flow and EMF control loops are used, where, accordingly, the P-regulator of flow and I-regulator of EMF are used. The current, speed, EMF and flow controllers (CR, SR, ER and FR) are tuned according to the modular optimum [2]. The advantage of this system is a standard, fairly simple (typical) method for calculating the parameters of the electric drive (regulators and sensors), the ease of setting up control loops, when the internal loop is first optimized and then the next in the structure, and the loops are set independently, it is quite simple to implement limitation of the main parameters of regulation.

Slave sequential regulation of coordinates allows you to provide almost any desired transfer function of the regulators on integrated operational amplifiers or in digital form [3, 4].

The disadvantages of typical systems of subordinate regulation are as follows. With a load surge, the output value of the speed controller is limited by the magnitude of the dynamic current reproduction error, which leads to underutilization of the electric motor power during acceleration under load and, accordingly, to a decrease in the productivity of the rolling stand. When working in the restriction zone at the input of the speed controller, a significant error accumulates, which, with a decrease in the static moment, decreases with some delay. At the same time, the drive becomes uncontrollable for a while. The disadvantages of the typical scheme can also be attributed to the engine EMF-dependent flow change for controlling the speed in the second zone. When working with large static moments, the motor voltage becomes limited, which leads to underutilization of the electric drive in terms of power, and at low static loads the motor is underused for anchor voltage.

IOP Conf. Series: Materials Science and Engineering **865** (2020) 012009 doi:10.1088/1757-899X/865/1/012009

To eliminate these drawbacks, an improved electric drive is proposed that contains a two-zone speed control system, where, unlike a typical system, an electric motor flow control independent of the EMF is implemented. The structural diagram of the improved automatic control system is shown in figure 1 and contains a dual-circuit system for controlling the EMF to the main speed of the electric motor and a dual-circuit system for regulating the flow above the main speed.

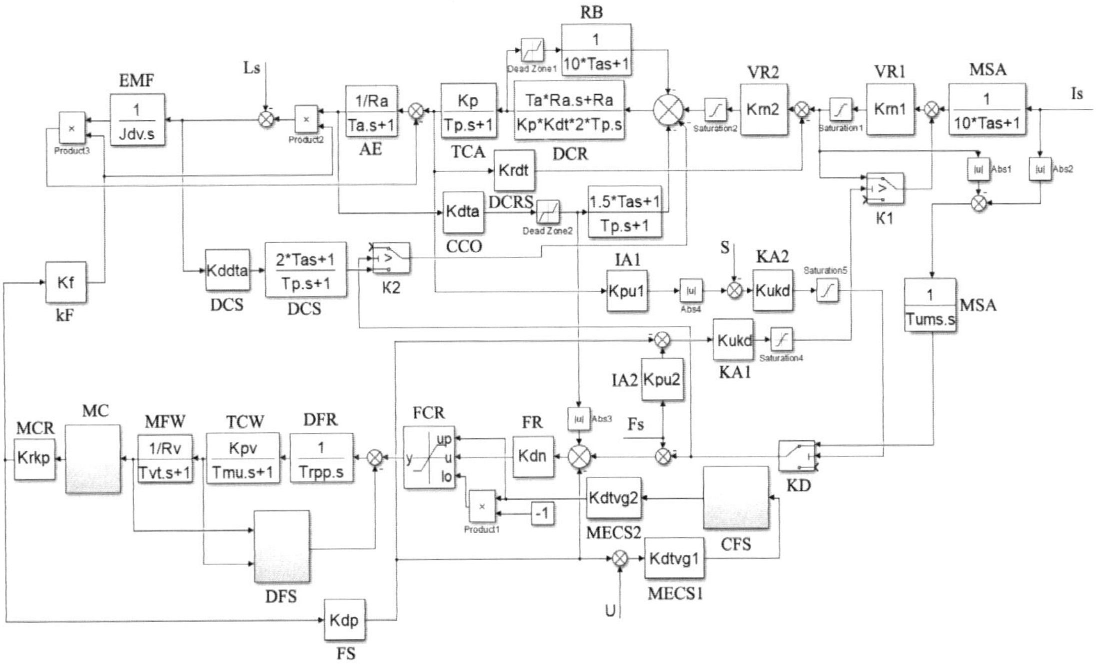

Figure 1. Structural diagram of an improved dual-zone automatic control system.

The EMF control system contains an internal dynamic current circuit consisting of a proportional-integral regulator DTR and a dynamic current sensor DCS, as well as an external EMF control circuit consisting of a proportional regulator VR2 and a sensor ES. The motor flow control system comprises an internal control loop of the derivative of the electric motor flow, consisting of a proportional controller DFR and a flow derivative sensor DFS, as well as an external flow circuit, consisting of a proportional RP flow controller and a DP flow sensor. To limit the maximum current, a current-limiting unit is introduced into the circuit, containing a current sensor for the motor armature CCO and a delay unit DZ2 with a current-limiting setting. In the area of operation of the electric drive below the main speed, the current limiting signal acts on the dynamic current controller, and in the area above the main speed, it affects the input of the flow controller FR. To maintain the constancy of the dynamic current in the control zone above the main speed, the driving signal to the flow is generated as a function of the cube of the flow. To this end, the FR limitation is performed in the FCRB block with an adjustable limiting voltage coming from the flow cube sensor CFS. To switch from the first to the second regulation zone and backward, electronic keys K1, K2, KD are used, which are controlled by comparison devices KA1 and KA2 [5, 6]. The derivative of the flow of the motor pF (1) is implemented in the sensor of the derivative of the flow DFS.

$$pF = U_f - I_f \cdot R_f \tag{1}$$

U_f – the voltage across the motor excitation winding (ATS); I_f – the current of the ATS; R_f – the resistance of the winding.

The parameters of the regulators and sensors are set using known methods of tuning subordinate control systems [2]. A feature of tuning the derivative flow loop is the following simplification: since the change in the excitation current I_f occurs much slower than the voltage U_f due to the large inductance of the field winding, feedback on the excitation current in dynamic modes can be neglected. Thus, in practice, the contour of the derivative stream is tuned as the MFW voltage circuit to a modular optimum. Current limitation parameters were determined using the Matlab and Simulink software.

3. Analysis

To analyze the typical and the improved control systems, their modeling was performed using Matlab and Simulink. ACS models are shown in figure 2 and figure 3. The resulting transients are shown in figures 4 and 5.

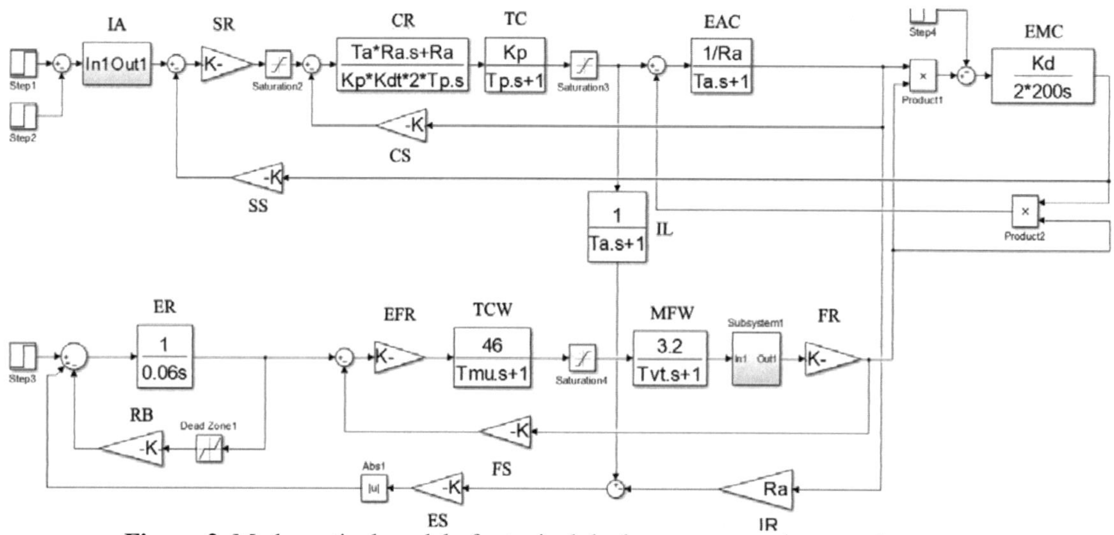

Figure 2. Mathematical model of a typical dual-zone automatic control system.

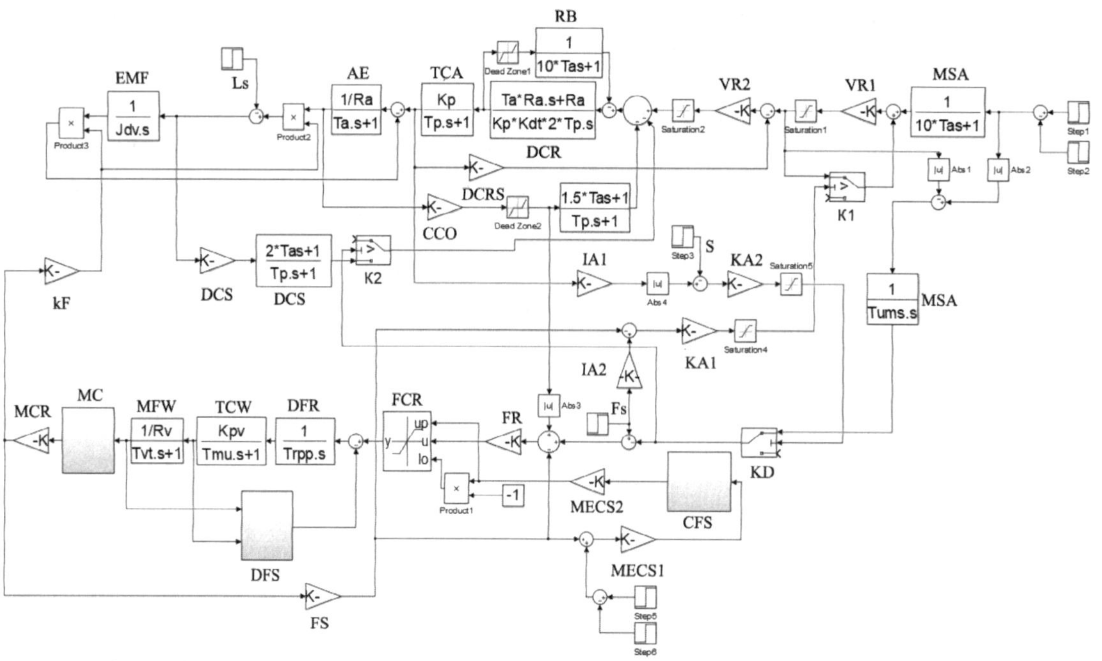

Figure 3. Mathematical model of an improved dual-zone automatic control system.

In a typical and improved electric drive system, in order to compare operating modes for different values of the static moment, the current limit setting was set to an armature current of 16 kA, and the acceleration speed of the electric drives at idle was also chosen to be the same - 4 rad / sec * sec.

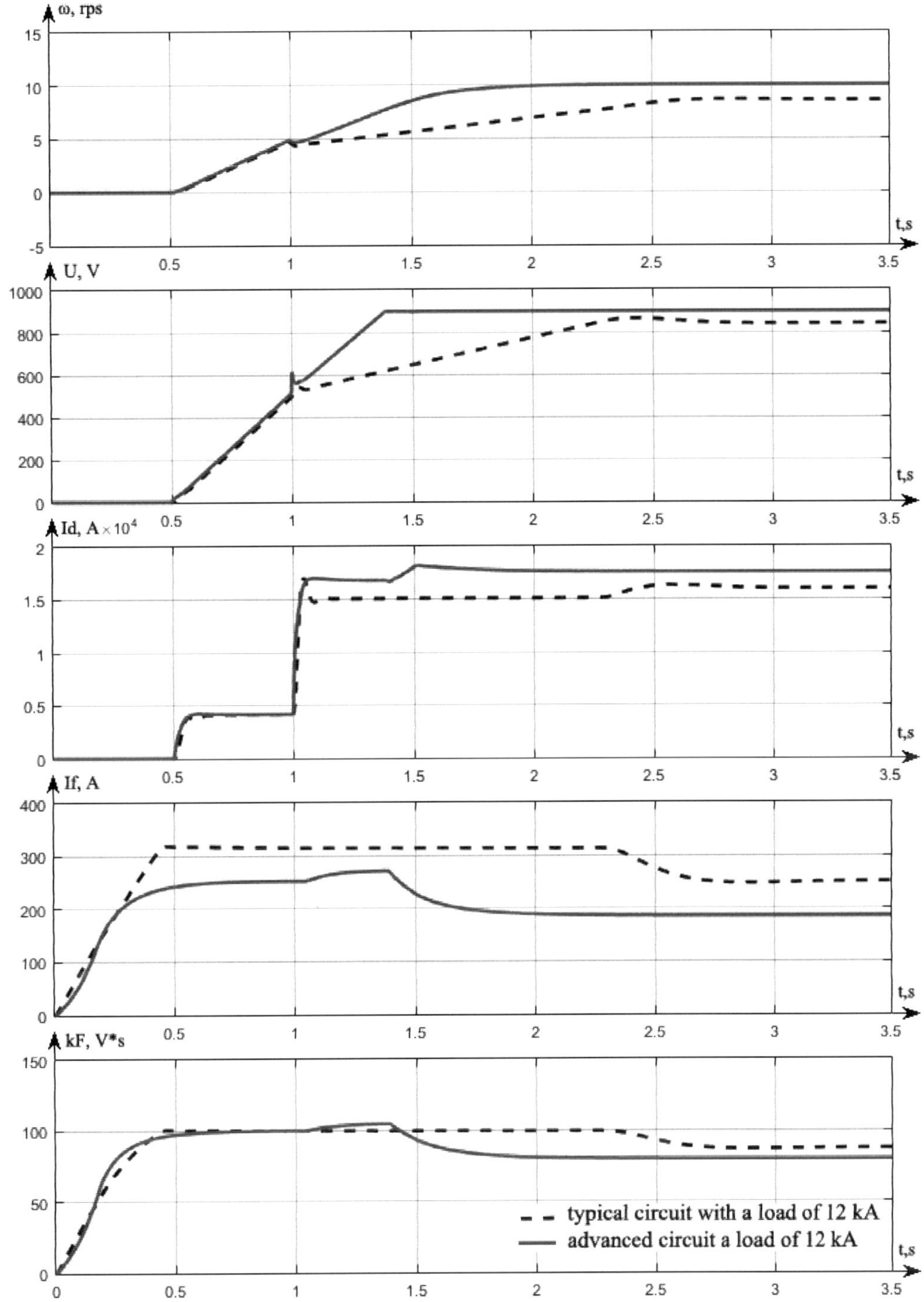

Figure 4. Transients obtained in the simulation of work under load.

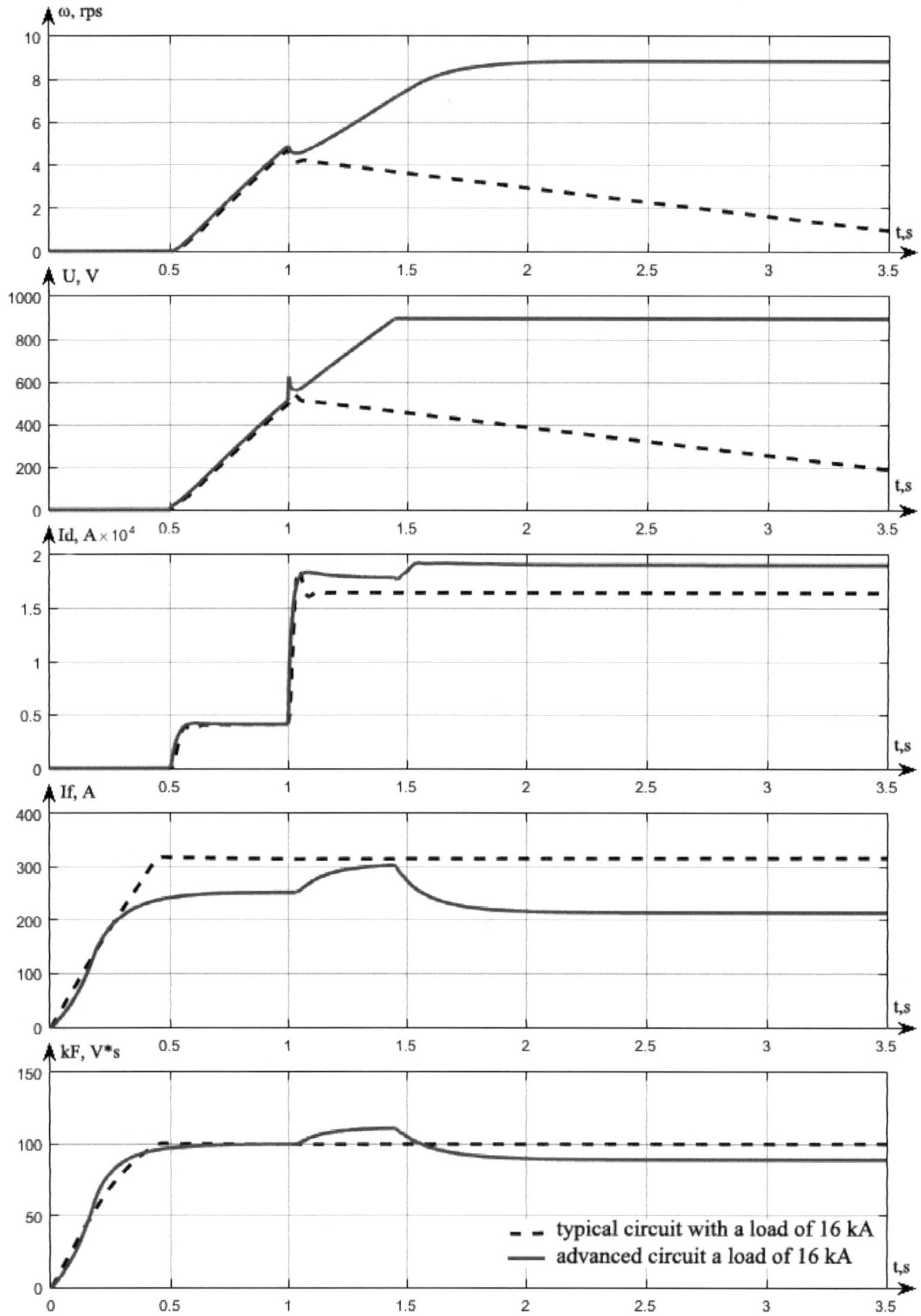

Figure 5. Transients obtained by modeling the operation of an electric drive with a maximum load.

4. Results and conclusions

From the analysis of transient waveforms shown in figure 4 it follows that at the moment of billet capture at a load current of 12 kA in a typical circuit, there is a decrease in the rate of increase in speed, which increases the acceleration time of the engine (the productivity of the mill decreases), while the rate of increase in the speed of the improved circuit does not change.

In figure 5 shows the parameters of electric drives with an increase in static load up to 16 kA. The rate of increase in speed in the improved dual-circuit design is reduced, since the armature current of the motor exceeds the current limit setting, while a typical dual-circuit system can no longer cope with such a load, as evidenced by the braking process until the motor stops and remains stationary under current, which may ultimately lead to an emergency.

Analysis of the operating modes of the electric drive with a typical and improved system of subordinate regulation revealed the following advantages of the improved system:

- In an improved control system during load surge, there is no drawdown of speed until the current limitation begins;
- High quality current limiting allows you to increase the maximum torque on the motor shaft;
- The motor of the improved system is fully used for anchor voltage.

The operating experience of the improved ACS in crimp and rail workshops of the Kuzmetkombinat showed the high quality of transients and reliable operation.

Referenses

[1] Belov M P, Novikov V A and Rassudov L N 2007 *Automated Electric Drives of Standard Research and Technological Complexes* (Moscow: Akademiya) p 376
[2] Muryshkin A K, Druzhilov S A and Bogdanovskaya T V 2007 *Calculation of a System of Subordinate Regulation by the Direct Current Electric Drive* (Novokuznetsk: SibSIU) p 30
[3] Terekhov V M and Osipov O I 2006 *Electric Drive Control Systems* (Moscow: Akademiya) p 304
[4] Garnov V K, Rabinovich V B and Vishnevetsky L M 1977 *Unified Systems of Automatic Control of Electric Drives in Metallurgy* (Moscow: Metallurgiya) p 102
[5] Kuznetsov V A 1983 *Sci. and Tech. Meeting on the Main Directions of Development of Modern Automated Electric Drive Systems and Issues of Operation and Repair of Electrical Equipment of Ferrous Metallurgy Enterprises* (Magnitogorsk) p 1– 35
[6] Kuznetsov V A 1981 *Pat. of the Russian Federation* No 972644 appl. 17.06.1981, publ. 07.11.1982

Functional status technological equipments modelling

G V Makarov, E V Tamarkina, M V Lyakhovets and A S Salamatin

Siberian State Industrial University, 42 Kirova street, Novokuznetsk, 654007, Russia

E-mail: gmakarov@nicsu.ru

Annotation. The paper deals with the modeling of the functional states of technological equipment that receive logical values "0" and "1" depending on the combination of input and output signals. The following functional states are possible: "Work", "Ready", "Not ready", "Fault". Model development these states and their minimization are possible by various methods, for example, with the help of Karnaugh maps, terminal machines, etc. The model variant for the simplest unit, obtained with the help of Karnaugh maps method, is presented. The application of this type of models is necessary for the typification of technological equipments, can be used to create logical control systems and integration into digital twins of complex production.

1. Introduction

It is accepted to describe the state of technological equipment by logical indicators, which can be divided into emergency and informing. Emergency indicators speak about equipment failure and impossibility of continuation (or the beginning) of work. Informing indicators serve for definition of a current condition of the equipment. Diagnostics of these indicators is carried out by system of local management on which information and operating functions are assigned. The logical control system should operatively estimate a condition of the equipment, form indicators, display the dispatcher and transfer in base for storage. In the event of an emergency indicator, the system must perform an algorithm of emergency stop or blocking the start. Informing indicators allow forming control actions to control the technological process.

The quality of logical control systems cannot be evaluated by the commonly used criteria of the theory of control – time of transients, stability, standard deviation, etc., as such notions are absent or do not make any sense for them.

The efficiency of a logical control system can be estimated by the time of reaction to emergency events, but mainly the direct relay protection of the launch equipment and additional protection devices are triggered. The system of logic management remains only to ascertain the fact, to generate the report and to deduce the corresponding warning to the operator - as it is done in modern automatic control systems at new designing and modernisation of existing [1-3].

In connection with the above, an indicator of the effectiveness of logical control systems (in addition, the correctness of the control algorithm and emergency interlocks) is the ratio of the number of diagnosed states to the total number of possible states of the unit or complex, which allow to quickly assess the situation, predict and prevent emergency states. In this case, the states can be "simple" (the presence / absence of a signal) and "composite" which depending on several logical or analog signals.

"Simple" states are the absence/presence of a signal, the simplest time conditions for the duration of triggering, acceleration or shutdown. "Complicated" states can be either in a specific unit or in a group of jointly operating mechanisms, such as a shut-off valve pump in the inlet, flushing and downstream. Component states of one unit may include indirectly evaluated states such as "slamming", overheating (when there is a continuous temperature measurement signal (or indirectly by thyristor temperature, by current evaluation of the drive temperature), but there is no protection in the form of a conventional thermal relay), the state of "heavy start", scrolling of the conveyor belt or "atypical" device readings, which may indicate an incorrect process or a malfunction of the device. Important indications are also the values of drive currents, pulsations and other characteristics. In combination, these can indicate the condition of the bearings, the drive itself and the need for maintenance. You can find an example of this diagnosis in [4]. Thus, controlling changes in parameters, it is possible to indirectly assess the condition of the equipment, for example, statistical methods, wavelet analysis, neural networks, the method of precedents [5], type-representative situations, etc., can be used.

To take into account all these states in the diagnostic algorithms of the local logical control system is quite difficult both for a novice developer (because of small experience) and an experienced one (because of the "soap" of the eye, or employment by other, more global and complex tasks, because such a developer is first of all tasked with developing an approach, or "system architecture"). At occurrence of changes in conditions of functioning of the unit, in its control algorithm, or in system architecture there is a task of adaptation of typical control systems to the changed conditions which, as a rule, falls on beginning developers.

For qualitative diagnostics it is necessary to take into account a great number of states of different complexity, both quite clear and not quite obvious. As an alternative, it can be done relying on existing programs or by interviewing a more experienced developer. But it is rather difficult to form a correct algorithm of processing and diagnostics of such states at once. Thought modeling often does not allow you to take into account all the peculiarities. And emergency modification of systems during their commissioning (commissioning) is accompanied by stress, errors, it takes a lot of time and may lead to deterioration of technological equipment. In this regard, it is necessary to use simulation complexes for so-called virtual commissioning.

For effective application of the approach to modeling of states of technological equipments first of all it is necessary to have a formal mathematical description and classification of states of technological equipments in the language of mathematical logic, supplemented for systems of real time – mathematical model of the object. Thus the set of diagnosed states is unique for each developer and is the intellectual property.

2. Model development

Let us describe possible models of equipment states on the example of the simplest control circuit – lamp switching on.

As an example in Figure 1 there is a circuit diagram of the lamp power supply circuit, in which KL1 is the control relay, KL2 is the circuit diagnostic relay, EL is the lamp, QF is the protective automatic switch (voltage supply), CV is the voltage control. The given circuit reflects all possible sets of possible signals - input controllers, status signals and a feedback control signal.

Figure 2 shows the circuit diagram to control switching the lamp on and off, corresponding to the circuit in figure 1. The figure also shows the logic variables for the input (S_1, S_2, S_3) and output (u) signals.

The functional status of the unit can be as follows: operation, ready for operation, not ready for operation and accident (F_1 ... F_4 respectively). For diagnostics, it is necessary to display not only the functional status of the unit, but also the causes.

IOP Conf. Series: Materials Science and Engineering **865** (2020) 012010 doi:10.1088/1757-899X/865/1/012010

The causes are determined by signals and reflect changes in the external environment in real time, i.e. they can appear and disappear, sometimes at high speeds. For fixing of the occurred event on the operator interface and storing in the base the processed logical variables, defined by signals or interrelation of signals – signs serve. The number of indications shall be sufficient to describe all the unique states of the system. For this purpose, it is necessary to have so many indications that the number of their N_P placements is greater than the number of signal placements. $N_{(u+S)}$:
$N_{(u+S)} = 2^k = 2^4 = 16$, where the number of signals $k = k_{in} + k_{out}$

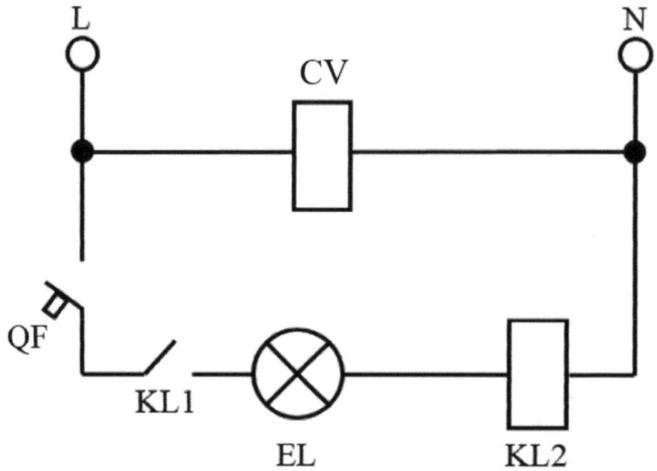

Figure 1. Circuit diagram for the room lighting system.

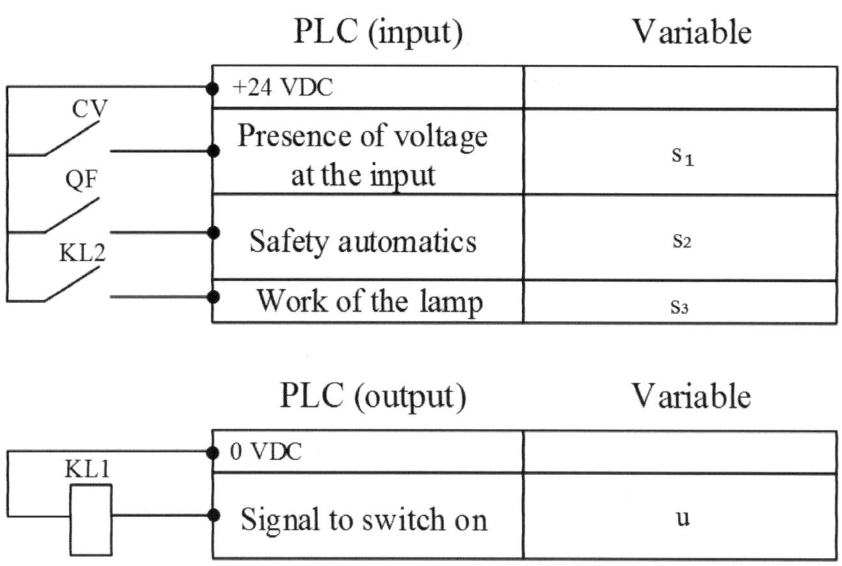

Figure 2. Circuit diagram of control.

The complete table of variants of combinations of logical input, output and variable functional states of the object looks like this (table 1).

Table 1. Summary table of input, output and variable states of the object.

Signals				Indications					Functional states			
Signal to switch on	Presence of voltage at the input	Safety automatics	Work of the lamp	There is no input voltage	The automatic machine is off (safety automatics triggered).	Lamp on	Lamp does not switch off (contactor fault)	Lamp malfunction (or chain fault)	Work	Ready	Not ready	Emergency
u	S_1	S_2	S_3	P_1	P_2	P_3	P_4	P_5	F_1	F_2	F_3	F_4
0	0	0	0	1	1	0	0	0	0	0	1	0
0	0	0	1	1	1	1	1	0	0	0	0	1
0	0	1	0	1	0	0	0	0	0	0	1	0
0	0	1	1	1	0	1	1	0	0	0	0	1
0	1	0	0	0	1	0	0	0	0	0	1	0
0	1	0	1	0	1	1	1	0	0	0	0	1
0	1	1	0	0	0	0	0	0	0	1	0	0
0	1	1	1	0	0	1	1	0	0	0	0	1
1	0	0	0	1	1	0	0	1	0	0	0	1
1	0	0	1	1	1	1	0	0	0	0	0	1
1	0	1	0	1	0	0	0	1	0	0	0	1
1	0	1	1	1	0	1	0	0	0	0	0	1
1	1	0	0	0	1	0	0	1	0	0	0	1
1	1	0	1	0	1	1	0	0	0	0	0	1
1	1	1	0	0	0	0	0	1	0	0	0	1
1	1	1	1	0	0	1	0	0	1	0	0	0

The methods of minimization for logical functions are well known – for example, the method of finite automata, Karnaugh maps, etc [6]. For a simple unit, each state can be visualized in the form of a Karnaugh map, after which we draw up tables of results, where u – inclusion KL1; S_1, S_2 and S_3 – states; $F_{1,2,3}$ – result. As an example, Figure 3 shows a Karnaugh map for the state of "Accident".

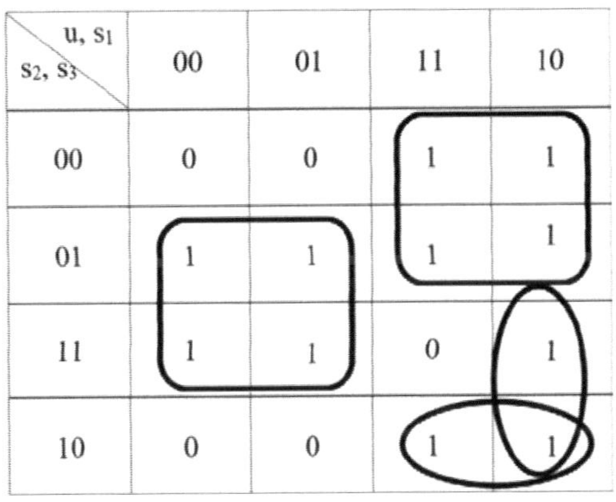

Figure 3. Karnaugh maps of the technological state of the "Crash".

Table 2 shows the function state models of F_1 ... F_4 in the language of logic algebra for all possible variants of signal placement and simplified expressions.

Table 2. Functional States

Variable	Model	Simplified model	Comment
F_1	$u \wedge s_1 \wedge s_2 \wedge s_3$	$u \wedge s_1 \wedge s_2 \wedge s_3$	Work
F_2	$\overline{u} \wedge s_1 \wedge s_2 \wedge \overline{s}_3$	$\overline{u} \wedge s_1 \wedge s_2 \wedge \overline{s}_3$	Ready
F_3	$\overline{u} \wedge \overline{s}_1 \wedge \overline{s}_2 \wedge \overline{s}_3$	$\overline{u} \wedge (\overline{s}_1 \vee \overline{s}_2) \wedge \overline{s}_3$	Not ready
	$\overline{u} \wedge \overline{s}_1 \wedge s_2 \wedge \overline{s}_3$		
	$\overline{u} \wedge s_1 \wedge \overline{s}_2 \wedge \overline{s}_3$		
F_4	$\overline{u} \wedge \overline{s}_1 \wedge \overline{s}_2 \wedge s_3$	$u \wedge (\overline{s}_1 \vee \overline{s}_2 \vee \overline{s}_3) \vee \overline{u} \wedge s_3$	Emergency
	$\overline{u} \wedge \overline{s}_1 \wedge s_2 \wedge s_3$		
	$\overline{u} \wedge s_1 \wedge \overline{s}_2 \wedge s_3$		
	$\overline{u} \wedge s_1 \wedge s_2 \wedge s_3$		
	$u \wedge \overline{s}_1 \wedge \overline{s}_2 \wedge \overline{s}_3$		
	$u \wedge \overline{s}_1 \wedge \overline{s}_2 \wedge s_3$		
	$u \wedge \overline{s}_1 \wedge s_2 \wedge \overline{s}_3$		
	$u \wedge \overline{s}_1 \wedge s_2 \wedge s_3$		
	$u \wedge s_1 \wedge \overline{s}_2 \wedge \overline{s}_3$		
	$u \wedge s_1 \wedge \overline{s}_2 \wedge s_3$		
	$u \wedge s_1 \wedge s_2 \wedge \overline{s}_3$		

3. Model simplification

Mathematical models in the language of logic are presented below.

$$F_1 = u \wedge s_1 \wedge s_2 \wedge s_3 \tag{1}$$

$$F_2 = \overline{u} \wedge s_1 \wedge s_2 \wedge \overline{s}_3 \tag{2}$$

$$F_3 = \begin{cases} \overline{u} \wedge \overline{s}_1 \wedge \overline{s}_2 \wedge \overline{s}_3 \\ \overline{u} \wedge \overline{s}_1 \wedge s_2 \wedge \overline{s}_3 \Rightarrow \overline{u} \wedge (\overline{s}_1 \vee \overline{s}_2) \wedge \overline{s}_3 \\ \overline{u} \wedge s_1 \wedge \overline{s}_2 \wedge \overline{s}_3 \end{cases} \tag{3}$$

$$F_4 = \begin{cases} \bar{u} \wedge \bar{s}_1 \wedge \bar{s}_2 \wedge s_3 \\ \bar{u} \wedge \bar{s}_1 \wedge s_2 \wedge s_3 \\ \bar{u} \wedge s_1 \wedge \bar{s}_2 \wedge s_3 \\ \bar{u} \wedge s_1 \wedge s_2 \wedge s_3 \\ u \wedge \bar{s}_1 \wedge \bar{s}_2 \wedge \bar{s}_3 \\ u \wedge \bar{s}_1 \wedge \bar{s}_2 \wedge s_3 \Rightarrow \begin{cases} u \wedge (\bar{s}_1 \vee \bar{s}_2 \vee \bar{s}_3) \\ \bar{u} \wedge s_3 \end{cases} = u \wedge (\bar{s}_1 \vee \bar{s}_2 \vee \bar{s}_3) \vee \bar{u} \wedge s_3 \\ u \wedge \bar{s}_1 \wedge s_2 \wedge \bar{s}_3 \\ u \wedge \bar{s}_1 \wedge s_2 \wedge s_3 \\ u \wedge s_1 \wedge \bar{s}_2 \wedge \bar{s}_3 \\ u \wedge s_1 \wedge \bar{s}_2 \wedge s_3 \\ u \wedge s_1 \wedge s_2 \wedge \bar{s}_3 \end{cases} \quad (4)$$

Thus, the complete model of a control object that takes into account all its possible states is described by expressions (1)-(4). Based on this model, a logical control system is developed, which allows to control the diagnosis of an industrial unit. The table of indications for diagnostics looks as follows.

Table 3. Status Models (indications).

Variable	Model	Type	Comment
P_1	$\bar{s_1}$	Emergency	There is no input voltage
P_2	$\bar{s_2}$	Emergency	The automatic machine is off (safety automatics triggered)
P_3	s_3	Informing	Lamp on
P_4	$\bar{u} \wedge s_3$	Emergency	Lamp does not switch off (contactor fault)
P_5	$u \wedge \bar{s_3}$	Emergency	Lamp malfunction (or chain fault)

4. Conclusion

When logical expressions are minimized, models are obtained that describe different technological functions and types of units. On this basis it is possible to make classification of typical objects for which quite certain typical systems of logic control are suitable - for example, the simplest fan, the pump, the screen [7] or more difficult, as, for example, the electric drive of the lift [8] will be described by one and the same model. The received logical models of an object and control systems allow to apply methods of similarity of control systems [9] as they allow to consider joint characteristics of an object and system, taking into account system features. Such typification at designing allows to reduce time of working out of the software and debugging at a stage of introduction of the automated control system.

References

[1] Shipunov M V, Korovin D E, Grachev V V, Myshlyaev L P, Lyakhovets M V, Melkozerov M Yu and Makarov G V 2018 *Vestnik of Siberian State Industrial University* **4(26)** 41–45
[2] Myshlyaev L P et al 2018 *IOP Conf. Ser.: Mater. Sci. Eng.* **354** 012008
[3] Ostrovlyanchik V Yu et al 2018 *IOP Conf. Ser.: Earth Environ. Sci.* **206** 012041

[4] Aniknanov D S 2016 *Introduction to ENERGYPTICS. Proc. of the II All-Russian (with international participation) Uouth Scientific-practical Conference (KuzSTU: Kemerovo) p* 206
[5] Kulakov S M, Musatova A I and Kadykov V N 2019 *Izvestiya Vuzov. Ferrous Metallurgy* **62** 652–659
[6] Chikurov N G 2003 *Logical Synthesis of Discrete Control Systems* (Ufa: Ufa State Aviation Technical University) p 132
[7] Salamatin A S, Makarov G V, Lyakhovets M V, Myshlyaev L P and Raskin M V 2018 *Science-intensive Technologies for Development and Use of the Mmineral Resources* **4** 330–337
[8] Ostrovlyanchik V Yu et al 2018 *IOP Conf. Ser.: Earth Environ. Sci.* **206** 012040
[9] Mishlyaev LP, Evtushenko V F et al 2012 *Steel in Translation* **42(12)** 823–841

The mechanism for the formation of a diversified production program of an enterprise, taking into account environmental and economic restrictions

T V Kiseleva[1], V G Mikhailov[2] and G S Mikhailov[2]

[1]Siberian State Industrial University, 42 Kirova str., Novokuznetsk, 654007, Russia
[2]T. F. Gorbachev Kuzbass State Technical University, 28 Vesennyaya str., Kemerovo, 650000, Russia

E-mail: kis@siu.sibsiu.ru

Abstract. Modern environmental requirements for the quality of products and technological processes, as well as toughening environmental legislation require improvement of the existing organizational and economic mechanism for environmental management. The article discusses the statement of the problem and the block diagram of the algorithm for the formation of the production program of the enterprise, taking into account environmental restrictions. Based on the actual data of an industrial enterprise with a diversified production program, the developed algorithm was tested, which showed the effectiveness of its use in practical activities.

1. Introduction

The effective operation of an enterprise depends on many factors, one of which is the adequate formation of a production program in an unstable environment. In [1], the production program is defined as an effective plan for the production of various types of products in kind and in cost terms in a timely manner, taking into account the requirements for its quality, nomenclature and assortment and developed on the basis of an analysis of market conditions and internal production capabilities of the enterprise using scientific management methods.

The study [2] is devoted to the development of a mechanism for the formation of the production program of an industrial enterprise in conditions of unstable demand for products using mathematical models. This approach to the management of the production program allows us to ensure maximum efficiency in the use of production capacities in conditions of market instability.

The tightening of environmental requirements [3-7] determines the need for the formation of the production program of the enterprise, taking into account environmental and economic restrictions.

2. Materials and methods

The object of study is industrial enterprises with a diversified production program that have a diverse negative impact on the environment. The subject of the study is the procedure for the formation of the production program. The study is based on an analysis of modern literary sources on the functioning of the existing organizational and economic mechanism for environmental management and its possible improvement. With the aim of practical implementation of the author's methodology, the developed algorithm was tested on the actual data of a large industrial enterprise. The work uses elements of a system analysis and the results obtained by experts in the field of environmental management.

3. Results and discussion

The idea of the proposed approach is that for enterprises with a diversified production program, it is important to plan the release of these types of products by item and assortment positions, which to a lesser extent have a negative impact on the environment and the ecological and economic result of the enterprise [8-11]. The main limitations when using this mechanism are related to the specifics of the production program, which should have a certain "degree of freedom" when choosing environmentally appropriate types of products. Other possible restrictions may include strategic production facilities, as well as the uniqueness of technological equipment that cannot be redirected or disadvantageously to other types of products.

Formulation of the problem.

Given:

1.The existing production program of the enterprise.

2.Nomenclature and assortment positions of enterprise products.

3.Profit from the sale of products of a specific nomenclature or assortment position.

4.Environmental and economic indicators of production of a particular item or assortment position $P_{S,T}$.

5.Restriction: $Pr_{S,T} > E^{*}_{S,T}$,

where $Pr_{S,T}$ – is the profit from the sale of products of a particular nomenclature or assortment position, $E^{*}_{S,T}$ – is the limit value of direct, indirect costs and risks associated with the elimination of the nomenclature or assortment position of the company's products.

6. The list of nomenclature or assortment positions of the enterprise's products, which can be replaced by ineffective ones.

7. Criterion: annual economic effect of the enterprise ($EE^{S,T}_{FPP}$), mln. rub.

$$EE^{S,T}_{FPP} = \begin{cases} Pr_{S,T} - E^{W}_{S,T} \\ Pr^{*}_{S,T} - E^{**}_{S,T} \end{cases}, \quad (1)$$

where $E^{W}_{S,T}$ – all types of costs associated with the production of a given nomenclature or assortment position ($PN_{S,T}$, $ED_{S,T}$, $F_{S,T}$ or $E^{*}_{S,T}$);

$Pr^{*}_{S,T}$ – profits from the sale of products replaced by an item or assortment position; $E^{**}_{S,T}$ – the costs associated with the replacement of the item or assortment position of products.

Required: to formulate the production program of the enterprise, subject to restrictions that maximize the criterion.

Figure 1 shows a block diagram of the algorithm for the formation of the production program of the enterprise, taking into account environmental and economic restrictions [12,13].

Figure 1 shows that in block 1, an analysis of the existing production program is performed. Block 2 shows a sequential review of each item or assortment position of the enterprise's products ($P_{S,T}$) for which environmental and economic assessment is carried out. As noted above, this is true for enterprises that produce a large number of types of products that differ, in addition, by grades, classes, concentration of the content of the main or useful substance, etc.

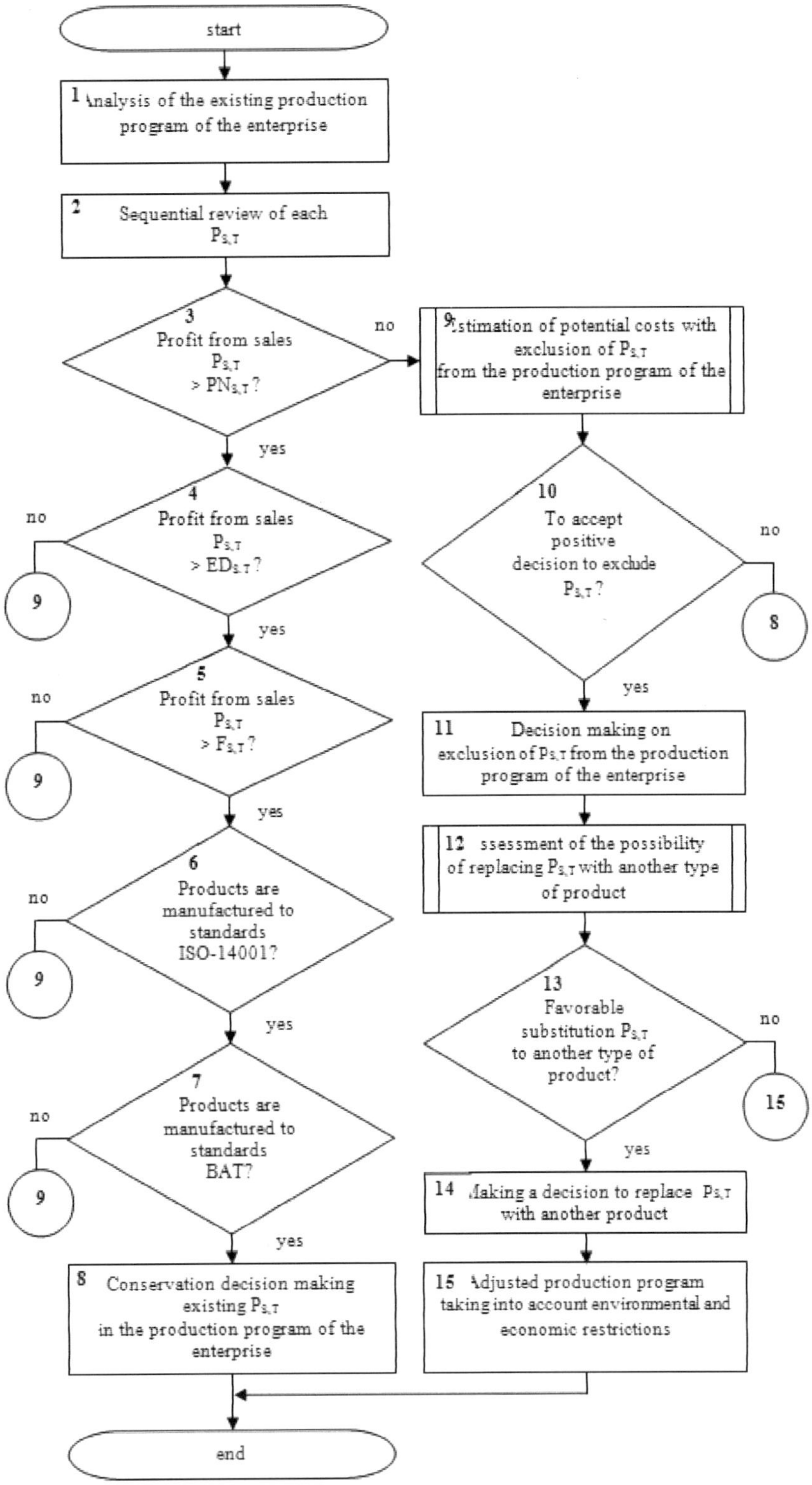

Figure 1. The block diagram of the algorithm for the formation of the production program of the enterprise, taking into account environmental and economic constraints.

IOP Conf. Series: Materials Science and Engineering **865** (2020) 012011 doi:10.1088/1757-899X/865/1/012011

Block 3 is necessary for comparing $Pr_{S,T}$ and fees for the negative environmental impact of a given $P_{S,T}$ ($PN_{S,T}$), which is very important in connection with the prospective increase in this payment. Similarly, $Pr_{S,T}$, is compared with the economic damage from environmental pollution ($EU_{S,T}$) and environmental fines ($F_{S,T}$) associated with the negative environmental impact of a particular $P_{S,T}$ – blocks 4, 5.

Calculation of $ED_{S,T}$, is carried out on the basis of well-known methods, including using the principles of EIA (environmental impact assessment). $F_{S,T}$ may be caused by a violation of environmental laws, the provision of false information about the negative impact on the environment and other factors.

Modern requirements for enterprises actively operating in the foreign market are associated with the need to use an environmental management system, including certification of products and processes according to the standards of the ISO-14001 series. In block 6, the compliance of this $P_{S,T}$ with the requirements of such a standard is checked. In logical block 7, the conformity of production to the criteria for using the best available technologies (BAT) by the enterprise to reduce the environmental burden on the environment is determined [14, 15], which is caused by modern environmental requirements, including the prospect of increased environmental payments.

In the case of a positive verification of $P_{S,T}$ on the basis of blocks 3-7, a decision is made to save it in the production program of the enterprise (block 8). If the results of the check in blocks 3-7 have a negative result, then in block 9, the direct, indirect costs and risks ($E_{S,T}$) associated with the liquidation of this $P_{S,T}$ are evaluated. The above restriction $Pr_{S,T} \geq E^{*}_{S,T}$ is determining in the statement of the problem of constructing an algorithm for the formation of the production program of the enterprise. The costs considered can be caused, for example, by the loss of a market niche or by the need to solve socio-economic problems due to the closure of individual industries and the release of workers. Based on the assessment, the final decision is made (block 10) to save the existing $P_{S,T}$ in the production program of the enterprise (block 8) or to exclude it (block 11). After that, in block 12, the possibility of replacing the excluded $P_{S,T}$ with another type of product is evaluated. In logical block 13, the "profitability" of substitution $P_{S,T}$ is determined. In case of a positive result, in block 14, a decision is made to replace the excluded PS, T with another type of product with the subsequent formation of the final production program of the enterprise taking into account environmental and economic restrictions (block 15). If the substitution $P_{S,T}$ is unprofitable, then the production program of the enterprise is formed in a "truncated" form.

Result of the decision. The presented flowchart of the control program for the production program taking into account environmental and economic restrictions can be implemented at enterprises with a diversified production program, when environmentally hazardous components are used in the production of certain types of products or their varieties, as a result of which the economic damage and payment for negative impact are significantly increased on the environment and, as a result, the likelihood of environmental and economic risks increases. This algorithm was originally developed for such a large chemical enterprise as the Kemerovo KAO Azot, which produces more than 40 types of products, but can be used in other sectors of the economy, taking into account the considered features of the production program.

Table 1 presents, as an example, several types of products of KAO Azot enterprise (caprolactam, POD oil, acidic caprolactam production (CASP), brand A ammonium nitrate, non-concentrated nitric acid of the 2nd grade), for which it is possible to accept decisions on the preservation, exclusion or replacement of a specific nomenclature or assortment position. The designations "+" or "-" indicate that the nomenclature or assortment position of products meets (does not meet) the established requirements. Profit from the sale of all considered types of products exceeds the values of fees ($PN_{S,T}$), economic damage ($ED_{S,T}$) and fines ($F_{S,T}$) for negative impact on the environment, and also complies with ISO-14001 environmental quality management standards.

Due to the fact that these types of products are produced without the use of BAT standards, in order to reduce or eliminate negative environmental impacts, a final decision is required on each of the considered item or assortment items.

Table 1 shows that caprolactam makes a decision on preservation in the production program, it is necessary to exclude POD oil and CASP, it is proposed to replace brand "A" ammonium nitrate with a new type of product (lime-ammonium nitrate) [16], and non-concentrated nitric acid 2nd grade partially replaced by another assortment position (nitric non-concentrated acid of the highest grade).

Table 1. An example of the implementation of the algorithm for the formation of the production program of the enterprise, taking into account environmental and economic constraints.

The name of the nomenclature or assortment position of the company's products	Profit from sales $P_{S,T}$ > $PN_{S,T}$?	Profit from sales $P_{S,T}$ > $ED_{S,T}$?	Profit from sales $P_{S,T}$ > $F_{S,T}$?	Products are manufactured to standards ISO-14001?	Products are manufactured to standards. BAT?	Making a final decision on the nomenclature or assortment position of the company's products (abandonment, exclusion, replacement)
Caprolactam	+	+	+	+	-	abandonment
POD oil	+	+	+	+	-	an exception
CASP	+	+	+	+	-	an exception
Ammonium nitrate brand A	+	+	+	+	-	partial replacement on lime-ammonium nitrate
Unconcentrated nitric acid 2nd grade	+	+	+	+	-	partial replacement non-concentrated nitric acid top grade

4. Conclusion

The study allows us to draw the following conclusions:

- various approaches to the formation and evaluation of the effectiveness of the production program of the enterprise, including the environmental and economic component; the statement of the problem of forming the production program is given taking into account environmental and economic constraints;
- a block diagram of the algorithm for the formation of the production program was developed taking into account the negative impact on the environment;
- analysis of well-known approaches to the distribution of environmental and economic resources, including the ongoing environmental costs in connection with their influence on reducing the level of environmental and economic risks, has been performed;
- the developed algorithm was tested, which showed that for caprolactam it is advisable to make a decision on preservation in the production program, and other products (AML oil, KSPK, brand A ammonium nitrate, non-concentrated nitric acid of the 2nd grade) should be replaced or excluded from production program;
- the developed algorithm for the production program formation is universal and has practical significance for enterprises of different industry sectors.

References

[1] Kushner A A 2010 *Bulletin of the Astrakhan State Tech. Univ. Series: Economics* **2** 89–94
[2] Ajupov A A, Kurilova A A et al 2016 *International Business Management* vol 10 **23** 5625–28
[3] Novichikhin A V, Fryanov V N et al 2017 *IOP Conf. Series: Earth and Env. Sci.* **84** 012036
[4] Tumin V M and Koryakov A G 2013 *Middle East J. of Sci. Research* 2013 vol 17 **9** 1350–55
[5] Zolotukhin V M, Gogolin V A et al 2017 *IOP Conf. Series: Earth and Env. Sci.* **50** 012027
[6] Zolotukhin V, Belkov A et al 2017 *E3S Web of Conf.* **15** 04015
[7] Tyulenev M A, Zhironkin S A et al 2017 *Geotech. Geol. Eng.* vol 35 **5** 2065–2077
[8] Burkov V N et al 2015 *Studies in Systems, Decision and Control* **10** 117–154
[9] Poryadina V, Burkov V and Barkalov S 2018 *MATEC Web of Conf.* **170** 01122
[10] Burkov V N et al 2015 *Studies in Systems, Decision and Control* 1-24
[11] Avdeev V P, Burkov V N and Kiseleva T V 2001 *Avtomatika i Telemekhanika* **3** 118–125
[12] Mikhailov V G, Kiseleva T V et al 2017 *IOP Conf. Ser.: Earth Environ. Sci.* **66** 012008
[13] Mikhailov V, Kiseleva T et al 2017 *E3S Web of Conf.* **21** 02004
[14] Makarevich E, Papin A et al 2017 *E3S Web of Conf.* **21** 02005
[15] Solodov V S, Cherkasova T G et al 2019 *Coke and Chemistry* vol 62 9 419–421
[16] Nabiev A A et al 2017 *Universum: Technical Science* vol 39 **6**

IOP Conf. Series: Materials Science and Engineering **865** (2020) 012012 doi:10.1088/1757-899X/865/1/012012

Flotation reagent dosing control system at the coil preparation plant LLC JV "Barsasskoye Tovarischestvo"

L P Myshlyaev[1], M V Lyakhovets[2], G V Makarov[1], D E Korovin[2] and G A Kulyushin[2]

[1]Scientific-Research Center for Control Systems" LLC, 55a Stroiteley ave., Novokuznetsk, 654005, Russia
[2]Siberian State Industrial University, 42 Kirova Street, Novokuznetsk, 654007, Russia

E-mail: mail@nicsu.ru

Abstract. The paper considers issues related to the creation of automated process control systems of the fleet-filtering department of the coil preparation plant LLC JV "Barzasskoye Tovarischestvo" (Berezovsky. Russia). The presented automated process control system is intended for controlling the flotation process and auxiliary operations. This system has been integrated into the control system of the entire factory and is intermediate between the main enrichment complex and the drying-furnace department. Integration is carried out at the technological and information levels. Effective management of the reagent dosing process allow to achieve minimal product losses.

1. Introduction

Coil preparation plants involved in the processing of minerals seek to increase production capacity constantly by introducing technological improvements in the structure of production. For example, technological processes that allow capturing small coal fractions, such as flotation [1]. The process of coal preparation by the method of flotation is effective in terms of quality indicators, but costly to use reagents. It is necessary to use the system of the automated control of technological processes that makes possible to use resources optimally - achievement of the best indicators of enrichment at the least expenses for manufacture.

As part of the next modernization at the coil preparation plant OOO JV "Barzasskoye Tovarischestvo" (Berezovskiy, Russia), it was decided to create a fleet-filtering department (FFD). The purpose of modernization was to increase the efficiency of enrichment processes of the technological complex of this department and, consequently, improve the technical and economic performance of the factory [2].

Coal sludge flotation is a process aimed at the enrichment of fine coal fraction. This process is based on selective adhesion of coal particles suspended in water to the surface of adjacent air bubbles. The most frequently used type of foam flotation, the mechanism of action of which consists in the following: pulp (a mixture of coal sludge with water) is mixed, and then it appears in the flotation machine, that implements the process of flotation. In the flotation machine, the pulp gets air consisting of tiny bubbles, the diameter of which does not exceed two millimeters. In addition, the pulp gets a flotation reagent, the presence of which does not allow air bubbles to collapse, getting into the pulp. The coal particles, having undergone preliminary treatment with the reagent, stick to the air bubbles and in such a condition appear on the surface of the pulp, subsequently forming the foam with enriched coal rock particles [3]. The use of flotation reagents in itself makes production more expensive, so it is necessary to dose them with high precision.

There are two approaches to solving this problem: empirical (manual) and automatic. In the first case, the operator independently regulates the supplied amount of floatation reagents filtrate. The quality of execution of this work directly depends on the operator's experience, which makes this method easy to use, but at the same time, the human factor influences the measurement of indicators. The second method assumes use of systems of automatic control (SAC, or SAR – system of automatic regulation) with application of concrete modes of operation of system that essentially raises accuracy of dispensing. However, this method also has significant drawbacks in the form of multiple input disturbances on the system, due to which the operating modes simply may be unstable at the set values. Consequently, it is necessary to create an automated control system that compensates for input disturbances.

2. Desctiption of the automation object of the fleet-filter section

At the coil preparation plant LLC JV "Barzasskoye Tovarischestvo" a fleet-filtering section object was realized, which is schematically shown in figures 1 and 2.

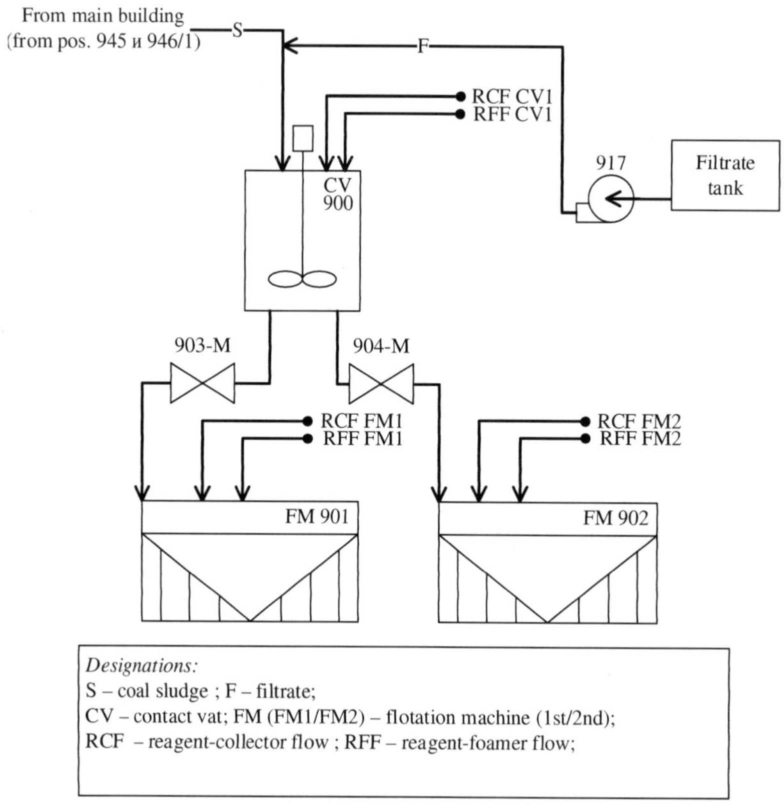

Figure 1. Flotation scheme.

Coal sludge (S) mixed with water comes from the main building by pumps 946/1 and 945, after that the resulting pulp enters the contact vat (CV) pos. 900, where the pulp is mixed with infeed flotation reagents: a collector (RCF CV1 flow) and a foamer (RFF CV1 flow) - by flotation reagent dosing systems (FRDS) 919/1 and 918/1 relatively. In addition, FRDS supplies flotation agents to chambers 3 and 5 of the flotation machines (FM). In figure 2 the designations, e.g. RCF FM1 C3, correspond to this, which means the flow of the reagent-collector to the third chamber of the flotation machine №1. The flotation concentrate received in CV is also sent to the flotation machines, where the enriched particles are separated from the waste rock, or the flotation tails, which are sent back to the main body to a radial thickener for a repeated cycle of the process. The flotation concentrate, in the form of foam, enters the mixing tank, where it is mixed with filtrate (F) formed by disk vacuum filters.

The concentrate is sent to the filters, by which the enriched product is separated from the filtrate and sent to the drying and furnace section by conveyor. The filtrate itself enters the filtrate tank (filtrate tank in figure 1), from which it is then fed by the 917 pump either back to the mixer or sent through a closed loop to the CV, or sent to the main building to radial thickeners.

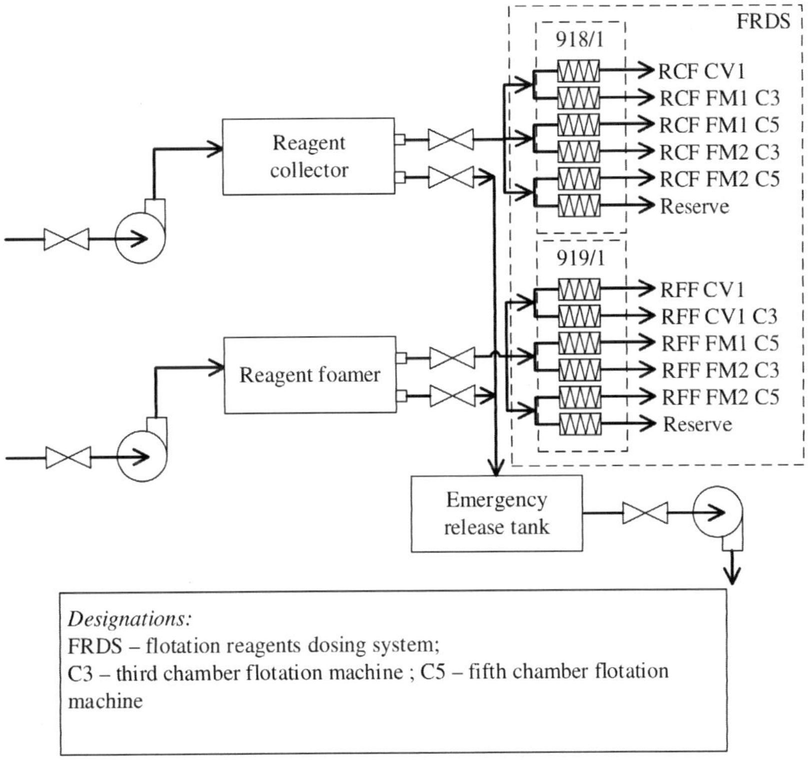

Figure 2. Scheme of the system for the preparation of flotation reagents

3. Model development

The task of the control system is to maintain reagent consumption depending on the operator's settings of the reagent to solid mass ratio in the pulp. For this purpose, flow and density are continuously measured:

1) density of products $P_L^{CV}(t)$ entering the contact vat (CV), where t – time;

2) quantity of product $Q^{CV}(t)$ entering the CV, is calculated on the basis of the set power of the pump pos. 946/1;

3) amount of filtrate $Q^F(t)$ coming from the pump (position 917) is also calculated by the pump power;

4) signal J^F about the infiltration of the filtrate into CV;

5) consumption of product $Q_F^1(t)$ from CV for the first flotation machine (FM1);

6) consumption of product $Q_F^2(t)$ from CV for the second flotation machine (FM2);

7) task on the ratio of the reagents supplied to CV, FM1 and FM2.

To the reagent-collector:

$$y_{RC}^{1*} = \frac{Q_{RCCV}^{1}}{Q_{RC}^{1}}; \quad (1)$$

$$y_{RC}^{2*} = \frac{Q_{RCCV}^{2}}{Q_{RC}^{2}}. \quad (2)$$

To the reagent-foamer:

$$y_{RF}^{1*} = \frac{Q_{RFCV}^{1}}{Q_{RF}^{1}}; \quad (3)$$

$$y_{RF}^{2*} = \frac{Q_{RFCV}^{2}}{Q_{RF}^{2}}. \quad (4)$$

where Q_{RC}^{1}, Q_{RC}^{2}, Q_{RF}^{1}, Q_{RF}^{2} – total amount of collector and foamer reagents on FM1 and FM2; Q_{RCCV}^{1}, Q_{RCCV}^{2}, Q_{RFCV}^{1}, Q_{RFCV}^{2} – the amount of collector and foamer reagents in CV during operation of FM1 and FM2;

8) signals J_{FM1} and J_{FM2} about the operation of FM1 and FM2, the physical analogues of which are represented by controlled valves 903-M and 904-M;

9) density of the filtrate P_{LF}^{CP}.

The structure of the control object is shown in figure 3.

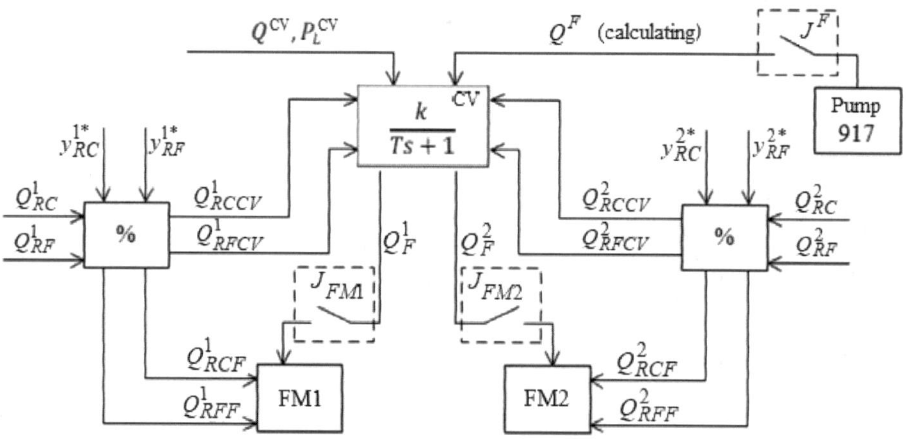

Figure 3. Structure of the control object of the dosing system.

Algorithm for calculating the dosage of the dosing system for the consumption of flotation reagents.

1. For the first FM:

a) for the collector reagent:
$$Q_{RC}^{1}(t) = k_{RC}^{1} * Q_{HF}^{1}(t); \quad (5)$$

$$Q_{HF}^{1}(t) = Q_{F}^{1}(t) * \hat{P}_{L}^{\sum CV}(t); \quad (6)$$

$$\hat{P}_{L}^{\sum CV}(t) = a_1 P_{L}^{\sum CV}(t) + a_2 P_{L}^{\sum CV}(t-1); \quad (7)$$

$$a_1 = \frac{T}{T+\Delta};\ a_2 = \frac{\Delta}{T+\Delta}; \tag{8}$$

$$P_L^{\sum CV}(t) = \frac{P_L^{CV}(t) * Q^{CV}(t)}{Q^{CV}(t) + Q^F(t)} + \frac{P_L^F(t) * Q^F(t)}{Q^{CV}(t) + Q^F(t)}; \tag{9}$$

b) for the foamer reagent:

$$Q_{RF}^1(t) = k_{RF}^1 * Q_{HF}^1(t). \tag{10}$$

2. *For the second FM:*
a) for the collector reagent

$$Q_{RC}^2(t) = k_{RC}^2 * Q_{HF}^2(t); \tag{11}$$

$$Q_{HF}^2(t) = Q_F^2(t) * \hat{P}_L^{\sum CV}(t); \tag{12}$$

$$\hat{P}_L^{\sum CV}(t) = a_1 P_L^{\sum CV}(t) + a_2 P_L^{\sum CV}(t-1); \tag{13}$$

$$a_1 = \frac{T}{T+\Delta};\ a_2 = \frac{\Delta}{T+\Delta}; \tag{14}$$

$$P_L^{\sum CV}(t) = \frac{P_L^{CV}(t) * Q^{CV}(t)}{Q^{CV}(t) + Q^F(t)} + \frac{P_L^F(t) * Q^F(t)}{Q^{CV}(t) + Q^F(t)}; \tag{15}$$

b) for the foamer reagent

$$Q_{RF}^1(t) = k_{RF}^1 * Q_{HF}^1(t). \tag{16}$$

Distribution of reagents between CV and FM.
1. *For the first FM*

$$Q_{RCCV}^1(t) = y_{RC}^{1*} * Q_{RC}^1(t); \tag{17}$$

$$Q_{RCF}^1(t) = (1 - y_{RC}^{1*}) * Q_{RC}^1(t); \tag{18}$$

$$Q_{RFCV}^1(t) = y_{RF}^{1*} * Q_{RF}^1(t); \tag{19}$$

$$Q_{RFF}^1(t) = (1 - y_{RF}^{1*}) * Q_{RF}^1(t). \tag{20}$$

2. *For the second FM*

$$Q_{RCCV}^2(t) = y_{RC}^{2*} * Q_{RC}^2(t); \tag{21}$$

$$Q_{RCF}^2(t) = (1 - y_{RC}^{2*}) * Q_{RC}^2(t); \tag{22}$$

$$Q_{RFCV}^2(t) = y_{RF}^{2*} * Q_{RF}^2(t); \tag{23}$$

$$Q_{RFF}^2(t) = (1 - y_{RF}^{2*}) * Q_{RF}^2(t). \tag{24}$$

Limitation check.
Q_{RC} и Q_{RF} on CV, FM1 и FM2

$$Q_{RC} < Q_{RC}^{\lim}, Q_{RF} < Q_{RF}^{\lim}. \tag{25}$$

During the modernization, the algorithm was implemented. Figure 4 shows a pop-up diagnostic video frame that is displayed when you click on the FRDS complex (918/1 and 919/1) on the main video frame. The window displays the elements of the general control of the FRDS (setting the operating mode, changing the task for supplying reagents), information on the status of the units FM1, FM2 and CV, control elements and data on the current operating modes of the pump group.

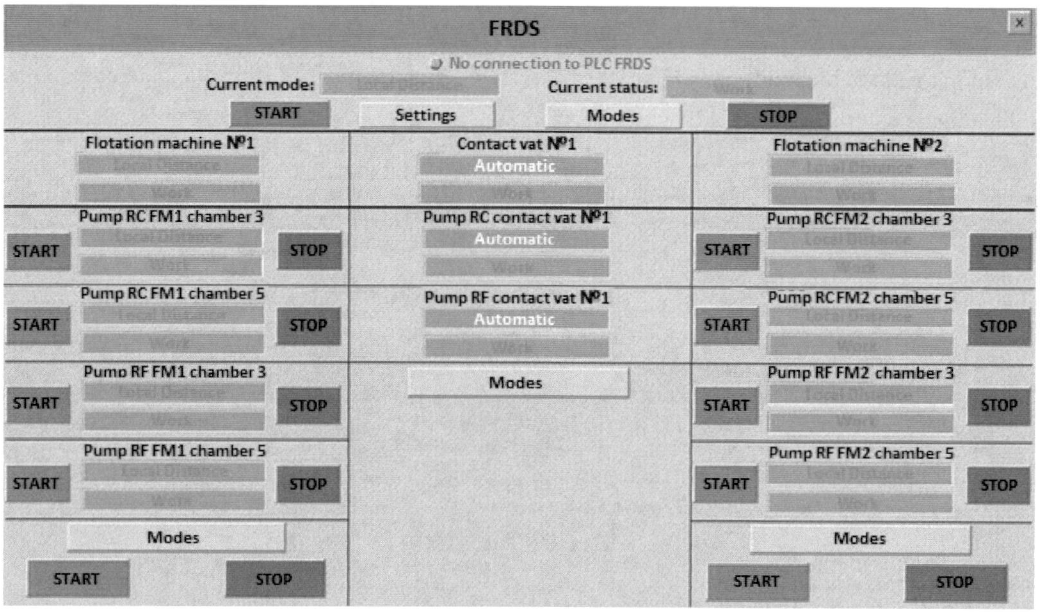

Figure 4. Diagnostic pop-up window of FRDS.

When you click on the area of a specific pump, a pop-up diagnostic window of the flotation reagent pump appears (figure 5). It contains information about the status and operating mode of the pump, fields for setting and displaying parameters and unit controls.

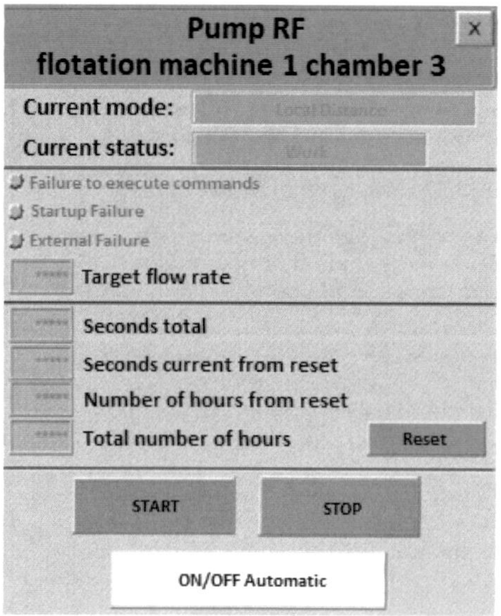

Figure 5. Diagnostic pop-up window of flotation machine 1 foamer reagent chamber 3.

IOP Conf. Series: Materials Science and Engineering **865** (2020) 012012 doi:10.1088/1757-899X/865/1/012012

4. Conclusion

The presented automated system involves the implementation of not only control functions, but also the functions of automated metering of pulp, solid matter in the pulp and the amount of spent reagents to be able to evaluate the overall performance of the system. The results of the study of the process, synthesis and tuning of the control system can be used as a standard solution [4] for other objects in compliance with their similarity [5].

References

[1] Lyakhovets M V, Venger K G et al 2017 *Proceedings of the XI All-Russian Scientific and Practical Conference in Automation Systems in Education, Science and Production* (Novokuznetsk: SibSIU) pp 151–156

[2] Myshlyaev L P, Makarov G V et al 2018 *Science-intensive Technologies for Development and Use of the Mineral Resources: Sci. J.* **4** 316–323

[3] Bogdanov O S, Podniek A K et al 1990 *Theory and Technology of Ore Flotation* (Moscow: Nedra) p 363

[4] Salamatin A S, Makarov G V et al 2018 *Science-intensive Technologies for Development and Use of Mineral Resources* **4** 330–333

[5] Myshlyaev L P, Evtushenko V F 2017 *Proc. of the XI All-Russian Sci. and Pract. Conf. in Automation Systems in Educ., Sci. and Prod.* (Novokuznetsk: SibSIU) pp 351–355

Creation of a control automation system on the example of the coal processing plant

M V Shipunov[1], V V Grachev[2], L P Myshlyaev[1], K A Ivushkin[3], Sh A Fayrushin[4] and G V Makarov[1]

[1]"Scientific-research Center for Control Systems" LLC, 55A Stroiteley str., Novokuznetsk, 654007, Russia
[2]Siberian State Industrial University, 42 Kirova str., Novokuznetsk, 654007, Russia
[3]Consolidated Company "Sibshakhtostroy" LLC, 9 Kuznetskoye highway, Novokuznetsk, 654034, Russia
[4]"Stroyservis" CJSC, 121 Kuznetskiy ave., Kemerovo, 650055, Russia

E-mail: mail@nicsu.ru

Abstract. Creation of automation system for control over technological processes of an preparation plant is considered. The functional structure, technical structure, hardware system solutions based on Hewlett-Packard, Omron and MOXA components, basic software based on Wonderware System Platform 2017 package are considered in detail. The selected hardware tools have reduced system development effort and increased system usability and efficiency.

1. Introduction

Coal mine No. 12 (Kiselevsk, Kemerovo Oblast) is the oldest existing coal mining enterprise in Russia. Since 1917, it has been developing the subsoil area of the south-eastern part of the Kiselevsk coal deposit. From the coal enterprise was born a working village, which became the city of Kiselevsk.

In 2013, it was decided to develop the area of open-pit mining and to stop the above-category hazardous underground mining. The annual production level of K and T grades of coal is over 730 thousand tons.

Currently, a new modern preparation plant (PP) of "Mine No. 12" is under construction with a production capacity of 4 million tonnes of raw coal per year.

Among the main design and technological solutions of PP "Mine No. 12" one can note [1-2]:

- application of coal preparation methods, which give low energy intensity of production;
- optimal combination of domestic and foreign equipment with its integration into an efficient and highly reliable structure;
- flexibility of the technological scheme providing the possibility to change the structure of the object, the choice and realization of rational schemes and modes of the enrichment process;
- ecological compatibility and safety of all production, reached by application of the closed warehouses of raw coals and concentrates, the closed cycle of turnover of water, exception of traditional processes of drying of a concentrate;
- high level of automation of units, technological processes and in general the whole production using modern software and hardware.

IOP Conf. Series: Materials Science and Engineering **865** (2020) 012013 doi:10.1088/1757-899X/865/1/012013

The main goal of creating a system of automatic control (ACS) of the "Mine No. 12" – improvement of management efficiency of the technological complex and, consequently, improvement of technical and economic indicators of its functioning:

- reduction of specific production costs and increase in the yield of a suitable product;
- improving the quality of marketable products.

The objective is achieved in the following areas:

1) automation of information and control functions when solving tasks:
- operative formation and analysis of information about changes in operation modes and condition of technological processes, units and equipment, electric and thermal energy consumption;
- operatively coordinated correction of tasks for the mode parameters of technological processes;
- operational implementation of control solutions and regulation of technological parameters;
- control, accounting and analysis of technological and production discipline violations, management efficiency;

2) increase of reliability of the control automation system of the technological complex, efficiency and quality of control and management due to:
- application of modern technical means, methods and algorithms of automatic control, analysis, diagnostics and control of technological processes and equipment;
- complex and detailed display of information about the state of equipment and units, changes in technological parameters, actions of operational personnel in the system;
- use of modern microprocessor controllers as a technical base at the lower level of the system, having high reliability, long operating time for failure, easy replacement of failed elements, extension of technical structure, modification of mathematical and software;
- minimizing the number of unreliable electromechanical devices in local logic control and management systems of individual mechanisms and units.

2. Enlarged functional structure

The enlarged functional structure of the united automated technological complex of PP "Mine No. 12" is shown in figure 1. It consists of the following main systems:

1) the system of automation of control of loading and warehouse complex (ACS LWC) of PP "Mine No. 12", including:
- system for centralised data collection, processing and storage;
- the system of engineering support of the LWC;
- the automated system of operative dispatching control (ASODC) of a boiler-house complex;
- ASODC loading complex;
- ASODC pumping station;
- ACS by boiler room complex, loading complex, coal storage;
- local ACS of treatment facilities, pumping station;

2) control automation system for the beneficiation plant of "Mine No. 12", including:
- system for centralized collection, processing and storage of PP information;
- production process control system;
- system of engineering support of preparation plant
- ASODC by technological complex of preparation plant (ASODC TC PP);
- ACS storage of raw coal;
- ACS by coal reception;
- ACS by the main building of the PP;
- ACS by the drying and furnace department;
- ACS wagon dumpster.

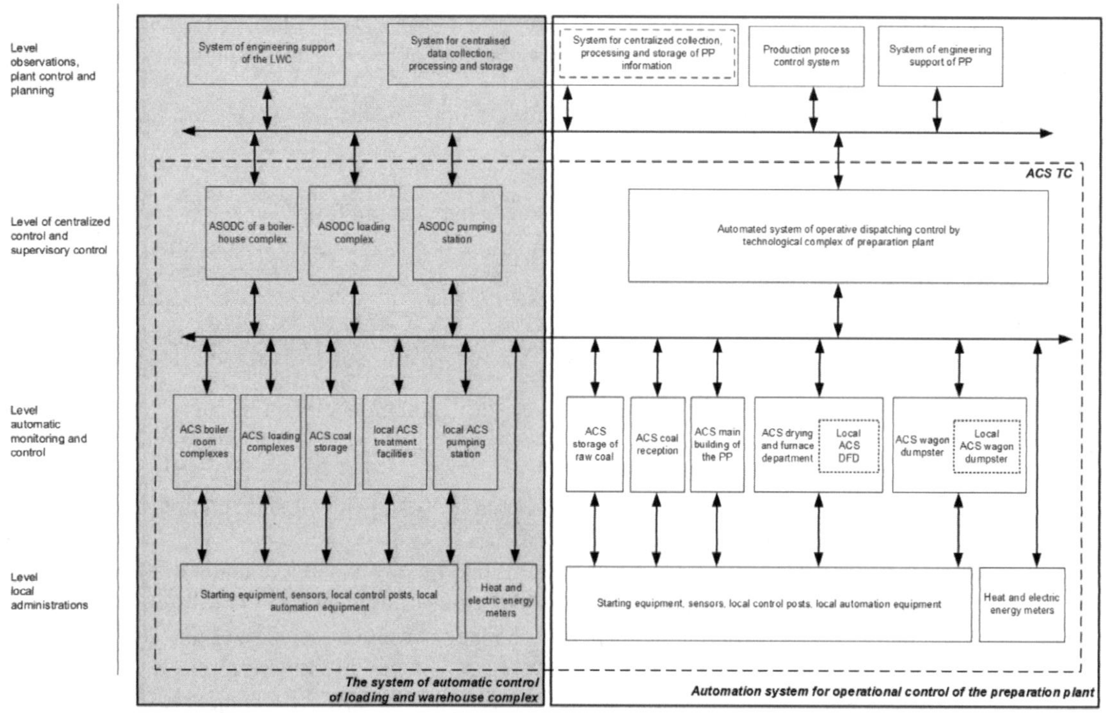

Figure 1. Scheme of the functional structure of the combined automated industrial complex of PP Mine No. 12.

Integration of the existing ACS LWC and the projected ACS TC PP and creation of the united automated technological complex of the PP is provided by [3-5]:

- creation of a single information space in the system of centralized collection, processing and storage of information and in the automated system of operative dispatching control of the factory technological complex, the general discipline of data visualization and dispatching control;
- inclusion of control algorithms for coordination of technological modes and mutual equipment interlocking at the interface points of the existing loading and storage complex and the projected technological complex of the processing plant into the structure of the mathematical support of the ACS;
- partial use of hardware-software means of ACS LWC for realization of functional ACS TC PP.

3. Extended technical structure

The schematic of the enlarged technical structure of ACS PP "Mine No. 12" is shown in figure 2. Hardware implementation of the system is based on microprocessor-based programmable controllers, servers, disk array, personal computers, HMI-terminals, laptop, printer, switches and interface converters.

This set of technical means is sufficient to perform all automated functions. The technical means of the ACS PP are equipped with standard interfaces and exchange protocols, which allows for their seamless integration with other systems if necessary.

4. Hardware implementation of ACS top level systems PP

Hardware realization of systems of top level ACS PP is executed on the basis of servers, personal computers, a laptop, the printer, a disk file and switches of corporation Hewlett-Packard (USA), HMI-terminals of corporation Omron (Japan) and the network equipment of company MOXA (Taiwan).

IOP Conf. Series: Materials Science and Engineering **865** (2020) 012013 doi:10.1088/1757-899X/865/1/012013

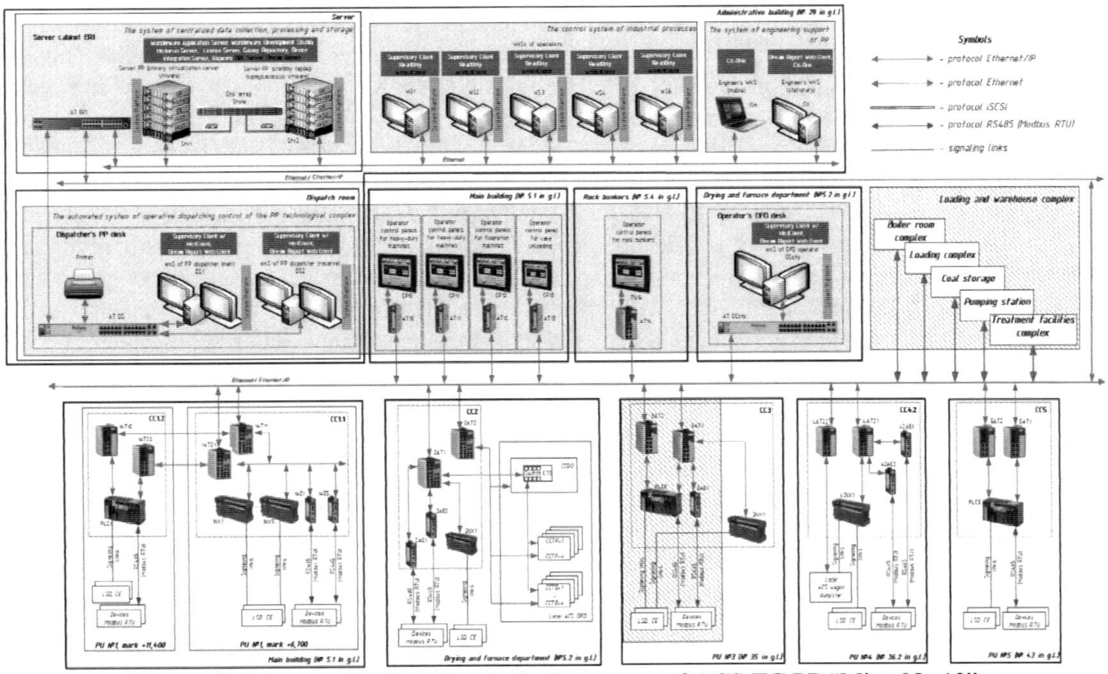

Figure 2. Scheme of enlarged technical structure of ACS TC PP "Mine No.12".

The system of centralized data collection, processing and storage consists of two physical virtualization servers (primary and backup) and a disk array. Each virtualization server is based on the HP Proliant DL360G10 server with two XEON processors. Each server houses five virtual machines:

- IAS1: Wonderware Application Server primary object server, Device Integration Server primary I/O server, Kepware OPC Server;
- IAS2: Wonderware Application Server standby object server, Device Integration Server standby I/O server, Kepware OPC Server;
- IDE: Wonderware Development Studio development server, Dream Report reporting software;
- History: Historian Server;
- GR: licensing server, Galaxy Repository.

For the initial configuration of the PP servers the HP LCD 8500 console and HPE 0 x 1 x 8 G3 KVM switch are used.

The disk array is implemented on the basis of HP MSA 1050 with 600 Gb hard drives combined into a RAID-array to exclude information loss in case of failure of one of the drives. The main function of the HP MSA 1050 drive array is to increase system resilience and flexibility.

During normal system operation, virtual servers (IAS1, IAS2, IDE, History, GR) are migrated to the disk array. The provider of data for workstations is the main virtualization server Srv1. In case of failure of the main virtualization server, its virtual machines are automatically migrated to the backup server Srv2, which is in the "hot" reserve. This option of redundancy provides increased fault tolerance of the system and data safety, eliminating the loss of production information.

The control system of industrial processes is realized in the form of WKSs of specialists of engineering and technical and planning economic services of the factory.

WKSs of specialists are implemented on the basis of personal computers HP ProDesk 400 G5 SFF with 27" LCD-monitors (5 WKSs in total) and are located in the AB, in the premises of the corresponding specialists of engineering, technical and planning and economic services of the factory.

IOP Conf. Series: Materials Science and Engineering **865** (2020) 012013 doi:10.1088/1757-899X/865/1/012013

The system of engineering support of PP is intended for instrumental support of existing and design of additional information and software for controllers, servers, workstations, HMI-terminals of operators in the process of search of rational technological regulations, adjustment, further development and modernization of ACS PP. The system is represented by a stationary engineer's workstation (personal computer), implemented on the basis of HP ProDesk 600 G4 with 32" LCD monitor and mobile engineer's workstation (laptop), implemented on the basis of HP 15-bs017ur with 15.6" LCD screen. The engineer's workstations are located in the AB, in the control system room.

The automated system of operative dispatching control of the PP technological complex is represented by the control desk of the PP dispatcher with two WKSs of the PP dispatcher (main and reserve) and the network printer, the control desk of the drying and furnace department operator with the WKS of the DFD operator, and also five control panels CP10-CP14 of operators.

Dispatcher's desk of PP is implemented on the basis of the standard desk of the firm “ConsErgo” of the 200 series, on which two WKSs of dispatcher's PP (main and reserve) and network printer are located, and also the equipment for operative dispatching control and management is mounted. Dispatcher's desk of PP is located in the room of the AB dispatcher's office.

Each of the dispatcher's workstations is based on a personal computer HP ProDesk 600 G4 with two 32" LCD monitors. As usual, the supplier of real-time production data for the dispatcher's workstations is the virtual server IAS1 of the physical server of the PP. In case of failure of the virtual server IAS1, there is an automatic switchover to the standby server IAS2, which is in the "hot" reserve, which eliminates downtime and loss of production information. Switching of dispatching workstations to the standby virtual server takes place without pauses or delays. In case of failure of the main workstation of the PP dispatcher, a backup server is activated, which is in the "hot" reserve.

The workshop operator's console is based on the Rittal console and is located in the drying operator's room. The workshop operator's console is equipped with a workshop operator's workstation, which is based on a HP ProDesk 600 G4 PC with two 32" LCD monitors, as well as control and monitoring equipment.

Operator control panels are based on Omron HMI terminals with touch screens.

Operator control panels for heavy-duty machines, floatation machines and rock bunkers are based on Omron's 15" NS HMI terminal series, which are installed in Rittal’s hermetically sealed operator desks and placed at the respective operator workplaces.

Operator control panel for cake unloading is based on an Omron HMI NS-series 10" HMI terminal installed in an airtight Rittal control unit at the respective operator's workstation.

5. Hardware implementation of low level ACS systems PP

Hardware realization of systems of the bottom level ACS PP is executed on the basis of controllers of series CJ2 of corporation Omron (Japan) and the network equipment of company MOXA (Taiwan).

Functionality and technical characteristics of these controllers fully correspond to the tasks of their application in ACS PP.

Lower level systems are built using controllers connected to each other via an information network via Ethernet/IP, with the transition to fiber optic cable for connecting the controllers located in other locations.

For connection of external devices and autonomous systems with RS485 interface (Modbus protocol) corresponding Omron communication modules are used.

The exchange of information between controllers, with devices connected via RS485 to the corresponding Omron controller communication modules, with service stations, with PP servers, with operator’s HMI-terminals is carried out via the information network via Ethernet and Ethernet/IP protocols.

When the CJ2 controller runs out of I/O channels, an NX-based remote I/O basket is added and connected to the controller via Ethernet/IP using the corresponding NX-EIC202 communication module.

6. Basic software

"Mine No. 12" was selected as the basic software for ACS:

- Wonderware System Platform package by Wonderware, a structural unit of Schneider Electric Corporation (France);
- VMware vSphere virtualization system (USA);
- modular OPS server Kepware OPC Server by Kepware (USA);
- Dream Report reporting software by Ocean Data Systems (Israel);
- CX-One software for configuration, configuration, programming and debugging of Omron automation equipment (Japan).
- The basic ACS PP software is modular, as open as possible and includes the following software products:
- I/O tools (Device Integration Server, Kepware OPC Server);
- system platform (Wonderware System Platform);
- virtualization system (VMware vSphere);
- Wonderware Historian Server (WMware vSphere); archiving and storage tools;
- Data analysis and reporting software (Wonderware Historian Client, Dream Report);
- Production data collection and processing software (Wonderware Application Server);
- data visualization software (Wonderware Supervisory Client);
- system development and maintenance software (Wonderware Development Studio, CX-One).

There are three modes of control of the technological scheme in the control system of PF "Mine No. 12":

- automatic control (the "Automat" mode is the main mode in which all information and control functions are automatically implemented);
- remote control (modes "Distance", "Individual Distance"), in which the system automatically implements all information functions, but the control of each unit (on/off) is performed by the dispatcher;
- local control (modes "Local", "Individual Local" – auxiliary, adjustment), in which the system automatically implements all information functions, and the control of each individual item of equipment is carried out by commands from local control posts.

Selection of the control mode of the technological complex "Automat", "Distance", "Local" is carried out with the help of the control panel of the dispatcher PP.

7. Conclusion

Designing of ACS of PP "Mine No. 12" is carried out in a very short time – 5-6 months, with a standard time of design of 9-12 months. It became possible, at the expense of parallel, simultaneous performance of some design works, application of typical design decisions and use of environments of the automated designing.

Acknowledgements

The work was performed according to State Regulations of the Ministry of Education and Science of the Russian Federation № 8.8611.2017/8.9.

References

[1] Sazykin G P et al 2003 *Designing and Construction of a New Generation of Coal Preparation Plants* (Novokuznetsk: SibSGIU) p 126
[2] Myshlyaev L P, Kiselev S F, Ivushkin A A et al 2003 *Control Automation of Coal Preparation Plant* (Novokuznetsk: SibSIU) p 304
[3] Grachev V V et al 2015 *Proc. of X All-Russian Sci. Conf. on Automation Systems in Education, Science and Production* (Novokuznetsk: SibSIU) pp 90–95
[4] Grachev V V and Shipunov M V 2014 *Proc. of VI All-Russian Sci. Conf on Automated Electric*

Drive and Industrial Electronics in the Metallurgical and Mining-fuel Industries (Novokuznetsk: SibSIU) pp 226–232

[5] Lyakhovets M V et al 2019 *Proc. of Int. Sci. Conf. on High Technology of Development and Use of Mineral Resources* (Novokuznetsk: SibSIU) pp 295–299

IOP Conf. Series: Materials Science and Engineering **865** (2020) 012014 doi:10.1088/1757-899X/865/1/012014

Geotechnological aspects of mining processes automation

S A Kizilov, S M Nikitenko and M S Nikitenko

Federal Research Center of Coal and Coal Chemistry, Siberian Branch of the Russian Academy of Sciences (FRC CCC SB RAS), 18 Sovetskiy Ave., Kemerovo, 650000, Russia
T.F. Gorbachev Kuzbass State Technical University, 28 Vesennyaya Street, Kemerovo, 650026, Russia

E-mail: sergkizilov@gmail.com

Abstract. The article is devoted to the geotechnologies' implementation for the mineral resources integrated development on the basis of modern software and hardware systems integrated into the mining enterprise general management system. In particular, the authors discuss a scientific idea, which consists in exploring the possibility of forming technical and technological requirements for the design of robotic systems to ensure the effective and safe release of coal under-roof stratum on the face conveyor based on the use of automation and robotics elements.

1. Introduction

In connection with the exhaustion of readily available mineral reserves, deposits of rich coal seams, as well as alluvial diamondiferous deposits, characterized by difficult mining and geological conditions, are involved in the development: depth increases, bedding prevails in thick seams. New conditions require appropriate advanced science-intensive technical and technological solutions that ensure safe and efficient production [1, 2]. Importance of implementation and the readiness of the transition of the mining industry to digital – automated and robotic "unmanned" technologies is noted at all levels in Russia and abroad, which is reflected in the policies of the international technical committee ISO / TC 82 "Mining" and the subcommittee ISO / TC82 / SC 8 "Advanced automated mining systems".

The need to develop components of "unmanned mines" is confirmed in the long-term program for the development of the Russian coal industry until 2030, approved by the Government of the Russian Federation Decree of June 21, 2014 No. 1099-r, where the main direction of technological development in relation to underground mining provides for the development and implementation of "unmanned" mineral mining systems based on integrated mechanization and automation [3]. Therefore, the most important direction of sustainable development of the coal industry is the implementation of geotechnologies for the integrated mineral resources' development on the basis of modern software and hardware systems integrated into the general management system of the mining enterprise.

2. Robotics design

Work on the design of robotics used in underground production began in the second half of the 70's of the 20th century with the development of destruction machines controlled remotely by Brokk [4, 5]. In the future, a number of different projects and prototypes of remotely controlled systems designed to help with tunneling operations appeared, a special place should be given to the self-propelled robot, the anchor installer, which appeared in the USA, developed by the Berkeley Institute [6]. Drilling and installation of anchors was the main direction of underground robotics in the late 70's and the first half of the 80's of the 20th century.

At the same time, research and development work began on the development of robotic systems for underground work in the USSR, the peak of which occurred in the second half of the 80's. One of the most unique was the project of a robotic complex of equipment for drilling and blasting operations, developed in the Donetsk branch of Giprougolavtomatizatsia, capable of performing up to 210 various operations [7]. In the future, work on the application and development of robots for working underground until the mid-2010's was mainly carried out in two directions – robots for conducting inspections of emergency facilities and semi-autonomous remote-controlled self-propelled loaders and trucks for underground mining of ore; in the second direction the Atlas Copco company development in this area should be highlighted, with its underground transport control system and semi-autonomous scooptram automation moving systems, now it is Epiroc company [8].

Work in the first direction with varying success is being carried out in all developed countries up to the present day; self-orientation systems of mobile robotic systems in underground workings are of the greatest interest. In this area, scientists from the USA take absolute leadership, developments are carried out by both large companies and institutes, it is worth noting the work carried out at the University of Alaska Fairbanks, where a team of scientists with the participation of Dr. Richard Weiss and Rajiv Ganguli developed and tested in practice self-orientation methods for flying and self-propelled drones in the mines [9]. Despite this, knowing the modern extracting equipment for mining, we see that complex robotic systems have not yet been widely adopted. Robots are used only in dozens of mines and are mainly associated with heading, fastening the face, transportation, and at enterprises that are not hazardous in gas and dust.

Since the mid-2010s, the German company Marco Systemanalyse und Entwicklung GmbH has been actively engaged in technological and engineering work on the introduction of elements, technologies, and methods of extraction work. So in the Kemerovo region by 2017, an experimental longwall was launched at the Polysaevskaya mine with the introduction of a robotic system for evaluating the longwall straightness, acoustic control of collisions on the roof, and monitoring the position of sections and their elements based on computer vision systems [10]. There are many companies on the world market that are developing modern systems for the automated control of mining equipment elements for a wide range of purposes. Companies developing and supplying integrated solutions for automated equipment management are an order of magnitude smaller.

The most advanced and famous are: Elgor + Hansen S.A. (Chorzow, Poland), FAMUR Group (Poland), MARCO Systemanalyse und Entwicklung GmbH (Germany), EEP Elektro-Elektronik Pranjic GmbH (Germany), Caterpillar (USA), TIEFENBACH Control Systems GmbH (Germany), Komatsu (Japan). Two companies – Caterpillar and Tiandi – present on their websites longwall complexes and related equipment for mining thick seams with free flow outlet of coal. However, the outlet technologies are similar to each other and coal outlet is carried out on a goaf conveyor belt, and not on a face conveyor [11]. This method of extraction has serious enough disadvantages that affect the complexity and cost of its use:

- a significant increase in the size of the support;
- the presence of an additional goaf conveyor;
- complication of the roof support structure;
- the need for the introduction of an additional reloading device into the complex;
- at the junction of the longwall with the conveyor gallery [12].

At the same time, the system of intelligent control of the outlet means (the protection shield on the goaf side, the goaf conveyor) is not considered as a separate subsystem, but is implemented through free (standby) channels on the basis of the produced automated electro-hydraulic control systems.

That is, a system for automated control of the outlet process is implemented on the basis of integrated longwall control systems; hydraulic control of technical means providing outlet (hydraulic jacks for opening windows and moving the conveyor) is carried out through additional (free) of electro-hydraulic control channels.

A promising platform for creating a new generation of robotic complexes for efficient and safe underground mining with the outlet of minerals from the overlying stratum and explosion-free methods of softening the massif is the fencing-supporting mechanized support with controlled outlet on the face conveyor developed by the FRC CCC SB RAS [13].

Technological features of the outlet process to the face conveyor, as well as the support sections' construction, does not allow the operator to effectively monitor the outlet process, which requires a transition from standard elements of the longwall complex automated control to robotic control systems with technical vision elements. The outlet process is included in the control system additionally, for this reason, the outlet technical means robotic control principles' development, even at the initial level, it is worth considering as the development of a control subsystem, which in the future is integrated into the applied system of complex intellectual and electrohydraulic control of the longwall complex.

When developing elements for the robotization of coal outlet to the face conveyor, one should focus on the possibility of integrating robotic outlet systems into a typical control system for sections of a longwall complex, which requires the use of standard hydraulic control elements and sensors from the main manufacturers of integrated mining automation systems. With this article authors' participation, technology simulation model for the effective development of coal deposits by a robotic complex with a controlled outlet of the under-roof stratum was developed at the FRC CCC SB RAS. The technology is evaluated and its rational parameters are determined, in which the use will be most effective. Using the developed tools, various modes of the under-roof stratum outlet were studied: individual, wave, group, and areal [14].

It was revealed that the optimal parameters cannot be maintained without the organization of complex systems for controlling the flow of rock from each section feeders, based on vision systems, which are the basis of the robotic support section with the under-roof stratum controlled outlet. To date, a prototype support section with controlled coal outlet to the face conveyor has been developed and is planned for production, based on RF patent No. 2513952 [15]. Mechanics and kinematics, simulation models are created, objects and methods for organizing the automated rock mass outlet from the under-roof stratum are described, without the outlet parameters and rock approach monitor systems' control possibility [16], however, the outlet process is not described in detail from the automation and robotization object point of view. The development technologies under different bedding conditions and types have some differences, however, the main tasks of automation and robotization are similar for all of them. For this reason, approaches to the design of control systems are largely similar.

The main tasks of automation and robotization in the framework of the rock mass production technological process include:

- composition determination ("rock-coal" boundary);
- conveyor loading assessment;
- an of the of each section output volumes estimation in real time;
- outlet initiation when the roof hangs, the of dome formation, blockage. The solution of the above problems has a number of serious limitations that do not allow the use of conventional methods used in other industries.

So, to evaluate the composition of incoming rock mass by standard methods is possible only when it is already on the main conveyor, and taking into account the output performance of each section on average about 50 kg/s and the presence of simultaneous outlet of, for example, 50 sections, up to 2.5 tons of waste rock can be on the conveyor in 1 second, which is a serious problem.

For optimal loading of the main conveyor, as shown by simulation, it is very important to maintain a given output from each section, but there is a problem here also. This is the use of a vibratory feeder for supplying the rock mass from the section window to the conveyor, as it is installed in a confined space, on which presses the entire mass of rock above the feeder's window, which makes it impossible to use standard weight sensors and other similar devices, all this is complicated by restricting explosion fireproof demands for the measuring device. At the moment, the project is considering the assessment of the vibrating feeder loading process using optical control systems for moving objects [17]. It is also difficult to observe the goaf part of the longwall, since it is possible for the non-collapsed roof to hang there, which, as model experiments show, can be prevented by changing the outlet algorithms at adjacent sections.

The scientific novelty of the research results consists in the development of a technique for robotic estimation of the volume and composition of rock mass when the section feeder lets it out; in obtaining the support section outlet process optimal modes' dependence on the volume and composition estimation data provided by means of technical vision; in substantiating the basic technological requirements to the design of complexes with release to the face conveyor robotics elements.

3. Conclusion

The obtained scientific results in the form of an IDEF0 scheme for the decomposition of the outlet process in key tasks (robotization: rock-coal, load amount, initiation); outlet control system operation algorithms; methods for assessing the volume and composition of the rock mass when the section is released by the feeder; technological requirements for the robotic complexes with controlled coal outlet creation; functional diagrams of the robotics of the support section feeder allowing, according to the authors, to solve the problem of the practical implementation of geotechnologies for the integrated development of mineral resources on the basis of promising software and hardware systems integrated into the general management system of the mining enterprise.

Acknowledgments

The reported study was funded by Russian Foundation for Basic Research (RFBR) according to the research project No. 19-35-90075\19.

References

[1] Goosen Ye V, Kagan E S et al 2020 *Smart Innovation, Systems and Technologies* pp 138–148
[2] Nikitenko S M and Goosen Ye V 2017 *Mineral resources of Russia. Economics and Management* **1** 27–32
[3] Order of the Government of the Russian Federation of June 24, 2014 No. 1099-r "Russia's coal industry development programs for the period until 2030"
[4] *Brokk history* https://www.brokk.com/about-brokk/history
[5] Thring M 1977 *UK Study: Remote – Controlled Miner. Coal Age* **6** 16–22
[6] Koniukh V L and Tailakov O V 1991 *Predesign Analysis of Mine Robotic Systems* (Novosibirsk: Nauka, Siberian Branch) pp 9–15
[7] Kiklevich Yu N 1987 *Mine Robotics* (Kiev: Tekhnika) pp 48–65
[8] *SME Mining Engineering Handbook* 2011 ed P Darling (Society for Mining, Metallurgy and Exploration Inc.) pp 808–812
[9] Gill P, Hatfield M et al 2015 *Conference: 51st AIAA/SAE/ASEE Joint Propulsion Conference* (Orlando) 4111
[10] Reuter M, Krach M et al 2017 *Fundamental and Applied Issues of Mining* No 2 Vol 4 263–269
[11] *The Principle of Longwall Top Coal Caving (LTCC)* https://www.youtube.com/channel/UCjYrT--TpOrA86Iw_PkoVJw?&ab_channel =CaterpillarGlobalMining
[12] Klishin, V I, Anferov B A and Kuznetsova L V 2017 *Innovations in the Fuel and Energy Complex and Mechanical Engineering (Fuel and Energy Complex-2017)* (Kemerovo: KuzSTU) pp 57–63
[13] Nikitenko M S and Kizilov S A 2019 *High-tech Technologies for the Development and Use of Mineral Resources* (Novokuznetsk: SibSIU) pp 257–263

[14] Starodubov A N, Zinoviev V V and Klishin V I 2019 *Science Intensive Technologies for the Development and Use of Mineral Resources* (Novokuznetsk: SibSIU) pp 253–257
[15] Klishin V I, Kokoulin D I et al 2012 *Patent for Invention* RUS 2513952 12.26.2012
[16] Kizilov S A , Nikolaev P I et al 2017 *Proc. of the XI All-Russian Sci. and Pract. Conf. in Automation Systems in Educ., Sci. and Prod.* (Novokuznetsk: SibSIU) pp 241–245
[17] Shubnikova I S and Plaguta K A 2013 *Proceedings of the Reliability and Quality International Symposium* (Penza: Penza State University) pp 352–355

IOP Conf. Series: Materials Science and Engineering **865** (2020) 012015 doi:10.1088/1757-899X/865/1/012015

Mirror symmetry in transverse oscillations of ropes and shafts in electromechanical systems

M Yu Borshchinsky

Siberian State Industrial University, 42 Kirova Str., Novokuznetsk, 654007, Russia

E-mail: bormax2005@mail.ru

Abstract: The article explores the possibility of developing a mathematical description of the bending of a round beam, based on end supports, to construct a theory of oscillations of a curved shaft of a mine hoist. The results of the study allow to develop a model of transverse oscillations of the elastic shaft and a long rope in order to synthesize a microcontroller control system.

1. Introduction

The transverse bending of the shaft is closest to that of the round beam (figure 1). But the beam rests freely on the supports, its ends bending upwards. Unlike the beam, the shaft is fixed in bearings, connected to other devices that prevent bending, which excludes the possibility to use theory [1, 2, 3, 4, 5] for the shaft.

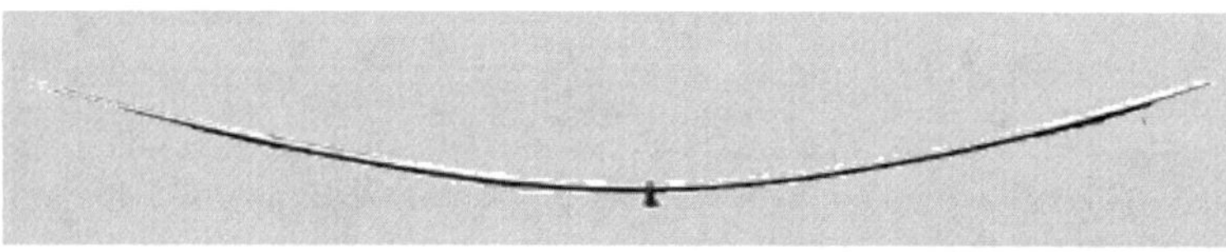

Figure 1. Experimental bending of a thin elastic round shaft.

The rope is pulled by a load and approaches the string in flexibility. However, according to Figure 2, the string is broken at the points of attachment and retraction, which is physically impossible for the rope. At the same time, the attachment of the rope ends is closer to the attachment of the shaft ends. So fastening the rope to the vessel with the load also does not allow a kink, and the rope runs onto the hoist drum and the pulley of the headframe tangentially to the circle, which physically excludes bending at the end point of tangency of the pulley circumference. It follows that the rope and shaft are subject to the same theory. However, the additional accounting of rope tension will increase the accuracy of the mathematical model.

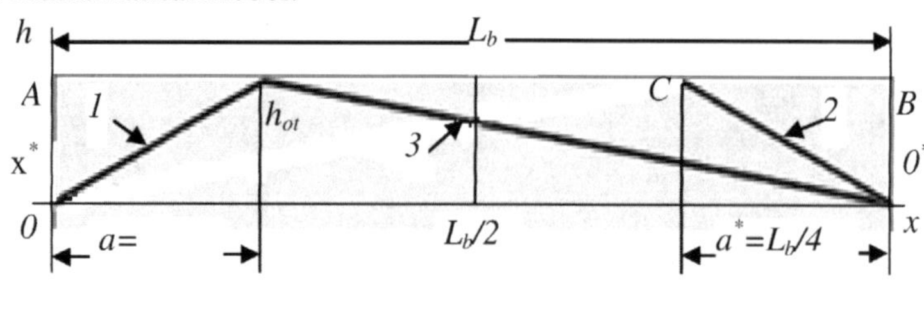

Figure 2. Bend string and mirror line 1 to line 2.

2. Results and discussion

Figure 3 shows the rope of the physical model of the mine hoist. At the pull point, the rope bends without a break, just like the shaft, but unlike a shaft with a smaller radius of curvature.

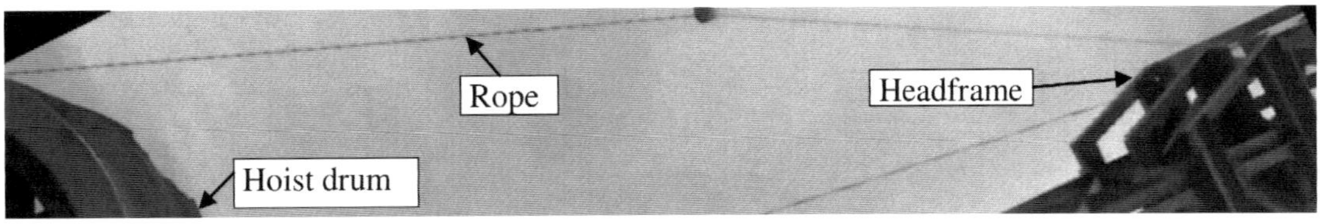

Figure 3. Rope, hoist drum and headframe of the laboratory stand of mine hoist.

The theory of bending in [4, 5] is built using bending moments M (Figure 4) applied to the beam ends, which made it possible to solve the problem of bending a freely supported beam, but it is impossible to fulfill the condition of pinching the shaft ends.

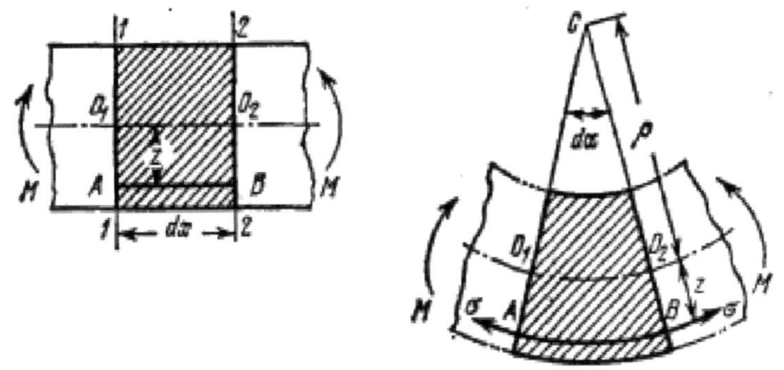

Figure 4. To the theory of bending of the beam freely supported at the ends.

The solution of this problem with respect to the shaft is based on Figure 5, which shows the concentrated force *P*, which is applied to the right face $x = a$ of the selected element of infinitely small length dx. The view of the element before and after bending is shown in Figure 5.

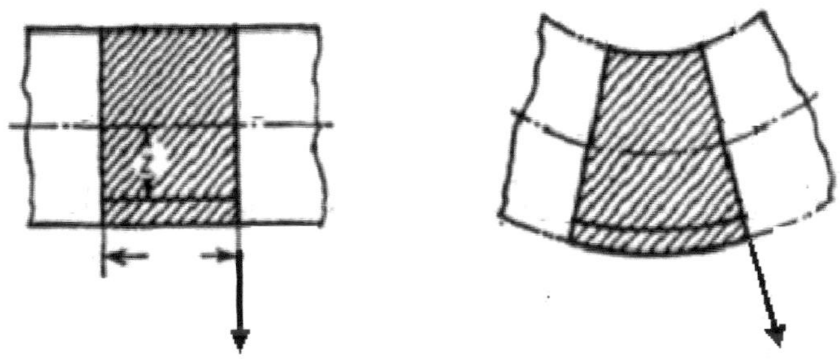

Figure 5. Bending of rope and shaft due to concentrated force.

Applying force *P* causes the shaft and rope to bend. When bent, the length of the axis O_1-O_2 (Figure 4) of the shaft and rope remains unchanged, and the fibres are compressed above the axis and stretched below. Compressive and stretching forces according to Hooke's law grow when bending increases, compensating for the bending action of force *P*, the bending process will stop only when the sum of compressive and stretching forces is equal to force *P*.

From Figure 5 we have $dM(x) = Pdx$ and integrating we get the moment-of-force function P $M(x) = Px + C_1$. But force P creates not one, but two moments with shoulders a and "$L_b - a$, and the distribution of force P for two moments is determined by the reaction of supports P_1 and P_2 (Figure 6), which are inversely proportional to the lengths of shoulders [1, 3, 8], in the consequence of which for the left support we have

$$\frac{P_1}{P} = \frac{L_b\text{-}a}{L_b},$$

from where

$$P_1 = \frac{P(L_b\text{-}a)}{L_b}.$$

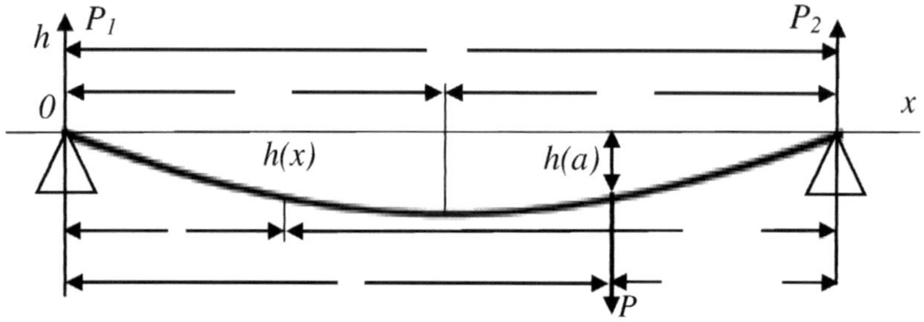

Figure 6. Design diagram of curved round beam.

By substituting P_1 in the moment function, we get

$$M(x) = \frac{P(L_b\text{-}a)}{L_b}x + C_1, \quad 0 \le x \le a. \tag{1}$$

The integration in Figure 6 extends only to the left side of the shaft, the solution of the problem with respect to the right side remains open. To solve the problem, we will use the same modes of fixing the shaft ends and ideas of mirror symmetry. The demonstration of the idea will be performed on a simple experiment with a string fixed on two supports A and B. Figure 2 shows the string in a pulled-off state and its mathematical description, borrowed from [1] has the form

$$h_1(x) = h_{ot} \cdot \frac{x}{a}, \qquad 0 \le x \le a \tag{2,a}$$

$$h_2(x) = h_{ot} \cdot \frac{L_b\text{-}x}{L_b\text{-}a}, \qquad a \le x \le L_b \tag{2,b}$$

In Figure 2, we introduce the second coordinate system, the origin $x^* = 0^*$ of which is located on the support B, and the x-axis x^* is directed from right to left, opposite the x-axis (we change the sign to the opposite) with the origin on the support 0. Based on the similarity of triangles we write

$$\frac{h_1(x)}{h_{ot}} = \frac{x}{a}, \qquad \frac{h_2(x)}{h_{ot}} = \frac{x^*}{a^*}, \qquad \text{откуда}$$

$$h_1(x) = h_{ot}\frac{x}{a}, \qquad 0 \le x \le a, \tag{3,a}$$

$$h_2(x^*)=h_{ot}\frac{x^*}{a^*}, \qquad a\leq x\leq L_b. \tag{3,b}$$

Figure 2 plots the functions (3, a) line 1 and the functions (3, b) - line 2. The plots are constructed in different coordinate systems and illustrate mirror symmetry, confirming the same operation modes of the supports. To construct the bending functions, the function (3, b) from the coordinate system x^* must be transformed into the system x.

According to [7] subject to conversion

$$x^*=L_b\text{-}x, \qquad a^*=L_b\text{-}a \tag{4}$$

as a result (3) will take the final form

$$h_1(x)= h_{ot}\cdot\frac{x}{a}, \qquad 0\leq x\leq a, \tag{5,a}$$

$$h_2(x)=h_{ot}\cdot\frac{L_b\text{-}x}{L_b\text{-}a}, \qquad a\leq x\leq L_b, \tag{5,b}$$

which corresponds to the graph of lines 1 and 3 in figure 2.

Using mirror symmetry, we write the bending function in the coordinate system x^*

$$M(x^*)=\frac{P(L_b\text{-}a^*)}{L_b}x^*+C_2, \qquad a\leq x\leq L_b. \tag{5,c}$$

Applying the transformation (4) to (5, c), we obtain

$$M(x)=\frac{P\cdot a}{L_b}(L_b\text{-}x)+C_2, \qquad a\leq x\leq L_b \tag{5,d}$$

Thus, for (1) and (5, d), we have

$$\mathrm{M(x)}=\begin{cases}\dfrac{P(L_b\text{-}a)}{L_b}x+C_1, & 0\leq x\leq a,\\ \text{-}\dfrac{P\cdot a}{L_b}x+P\cdot a+C_2, & a\leq x\leq L_b.\end{cases} \tag{6}$$

In sources [4, 5], only bending moments are taken into account, taking into account the additional dynamic moment and the moment of internal viscous friction $M_t(t)$, we obtain

$$EJ\frac{d^2h(x,t)}{dx^2}=M(x)+M(t)\text{-}M_c(t)\text{-}M_t(t), \tag{7}$$

where according to the source [8] we have

$$M(t)=\begin{cases}(L_b\text{-}a)m\dfrac{d^2h(t)}{dt^2}, & 0\le x\le a,\\ am\dfrac{d^2h(t)}{dt^2}, & a\le x\le L_b.\end{cases} \tag{8}$$

$$M_t(t)=C_t\,L_b\eta D\frac{dh(t)}{dt},\qquad C_t\,L_b=c_t \tag{9}$$

By substituting (6), (8) and (9) in (7), we obtain a differential equation in the second-order partial derivatives describing the damping oscillations of the shaft and rope

$$EJ\frac{\partial^2 h(x,t)}{\partial x^2}\text{-}Am\frac{\partial^2 h_1(x,t)}{\partial t^2}\text{-}B\frac{\partial h_1(x,t)}{\partial t}=\begin{cases}\dfrac{P(L_b\text{-}a)}{L_b}x+C_1,\ A=L_b\text{-}a & 0\le x\le a,\\ \text{-}\dfrac{P\cdot a}{L_b}x+Pa+C_2,\ A=a, & a\le x\le L_b.\end{cases} \tag{10}$$

Constant integration is determined from steady-state mode, when

$$\frac{\partial^2 h(x,t)}{dt^2}=\frac{\partial h(x,t)}{dt}=0,$$

in this case, the partial differential equation (10) is transformed into an ordinary differential equation

$$EJ\frac{d^2h(x,t)}{dx^2}=\begin{cases}\dfrac{P(L_b\text{-}a)}{L_b}x+C_1, & 0\le x\le a,\\ \dfrac{Pa}{L_b}(L_b\text{-}x)+C_2, & a\le x\le L_b.\end{cases} \tag{11}$$

Integrating twice, we get

$$EJ\frac{dh(x,t)}{dx}=\begin{cases}\dfrac{P(L_b\text{-}a)}{2L_b}x^2+C_1x+C_3, & 0\le x\le a,\\ \text{-}\dfrac{Pa}{2L_b}(L_b\text{-}x)^2\text{-}C_2(L_b\text{-}x)+C_4, & a\le x\le L_b.\end{cases}$$

$$EJh(x)=\begin{cases}\dfrac{P(L_b\text{-}a)}{6L_b}x^3+C_1\dfrac{x^2}{2}+C_3x+C_5, & 0\le x\le a,\\ \dfrac{P\cdot a}{6L_b}(L_b\text{-}x)^3+C_2\dfrac{(L_b\text{-}x)^2}{2}\text{-}C_4(L_b\text{-}x)+C_6, & a\le x\le L_b.\end{cases}$$

Fulfilling the conditions for fixing the ends of the shaft in the bearings, we obtain the boundary conditions

$$h(0)=h'(0)=h(L_b)=h'(L_b)=0 \tag{12}$$

we get $C_3=C_4=C_5=C_6=0$, the substitution of which gives

$$EJ\frac{dh(x,t)}{dx}=\begin{cases}\frac{P(L_b-a)}{2L_b}x^2+C_1x, & 0\le x\le a,\\ -\frac{P\cdot a}{2L_b}(L_b-x)^2-C_2(L_b-x). & a\le x\le L_b.\end{cases}\tag{13}$$

$$EJh(x)=\begin{cases}\frac{P(L_b-a)}{6L_b}x^3+C_1\frac{x^2}{2}, & 0\le x\le a,\\ \frac{P\cdot a}{6L_b}(L_b-x)^3+C_2\frac{(L_b-x)^2}{2}, & a\le x\le L_b.\end{cases}\tag{14}$$

The integration constants C_1 and C_2 are determined from the condition of continuity and smoothness of the function. For this, the first function from (14) must be equated with the second and the first derivative from (13) and the second with $x = a$. Then, grouping the terms in the obtained equalities, we have

$$C_1a+C_2(L_b-a)=-\frac{Pa^2(L_b-a)}{2L_b}-\frac{P\cdot a(L_b-a)^2}{2L_b},$$

$$C_1\frac{a^2}{2}-C_2\frac{(L_b-a)^2}{2}=\frac{P\cdot a(L_b-a)^3}{6L_b}-\frac{Pa^3(L_b-a)}{6L_b}.$$

We multiply the first equation by $\frac{L_b-a}{2}$, add it to the second, and solve the result with respect to C_1

$$C_1=\frac{P\left(a^2-L_b^2\right)}{6L_b}.\tag{15,a}$$

The first equation is multiplied by $\frac{a}{2}$ and subtract the second from it, resolving with respect to C_2, we obtain

$$C_2=\frac{P\cdot a\left(3aL_b-a^2-2aL_b^2\right)}{6\left(L_b^2-aL_b\right)}\tag{15,b}$$

3. Conclusion

The simulation result is shown in Figure 7. The curves were recorded using the support and research system developed by the author. Figure 7 a) shows the bending of the shaft from the action of the force P applied to the middle of the shaft $a=\frac{L_b}{2}$, n figures 7 b) and 7 c) the force P was applied to the points $a=\frac{L_b}{4}$ and $a=\frac{3L_b}{4}$.

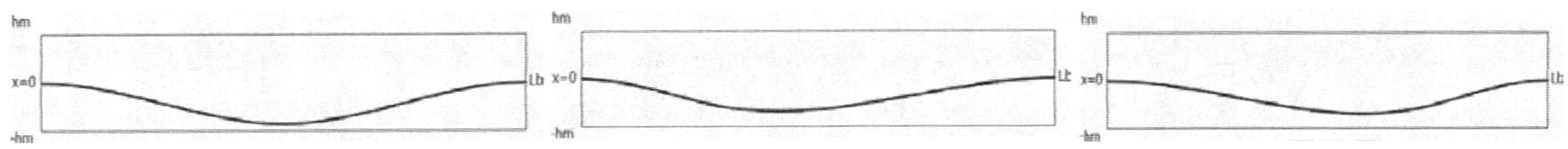

a) b) c)

Figure 7. Simulation results of shaft bending from the force P acting at the point $a=\frac{L_b}{2}$ for a), at the point $a=\frac{L_b}{4}$ for b) and at the point$a=\frac{3L_b}{4}$ for c).

Figure 8 shows the experimental curves, and figure 8 a) corresponds to figure 7 a), figure 8 b) to figure 7 b), and figure 8 c) to figure 7 c). A comparison of the figures shows the coincidence of the characteristics of theoretical processes with experimental ones.

a) b) c)

Figure 8. Experimentally obtained bending of the shaft from the force P acting at the point $a=\frac{L_b}{2}$ for a), at the point $a=\frac{L_b}{4}$ for b) and at the point$a=\frac{3L_b}{4}$ for c).

Thus, based on mirror symmetry using two coordinate systems, a universal mathematical model of the transverse oscillations of the shaft and the rope from the action of a concentrated load is created. The model describes forced and free oscillations, as well as a steady state. Accounting for internal viscous friction provides an investigation of damped oscillations.

References

[1] Koshlyakov N S 1970 *Equations in Partial Derivatives of Mathematical Physics* (Moscow: Vysshaya Shkola) p 712

[2] Piskunov N S 2001 *Differential and Integral Calculus* (Moscow: Integral-Press) vol 2, p 544

[3] Farlow S 1985 *Partial Derivative Equations for Scientists and Engineers* (Moscow: Mir) p 384

[4] Feodosiev V I 2000 *Resistance of Materials* (Moscow: MSTU named after NE Bauman) p 592

[5] Belyaev N M 1976 *Resistance of Materials* (Moscow: Nauka) p 608

[6] Landsberg G S 2010 *Elementary Physics Textbook. Mechanics. Heat. Molecular Physics* (Moscow: FIZMATLIT) vol 1, p 612

[7] Privalov I I 1966 *Analytical Geometry* (Moscow: Nauka) p 272

[8] *General Physics Course. Mechanics* https://phys.bspu.by/static/um/phys/meh/1mehanika

Models and mechanisms for planning service improvements

A V Zymin[1], I A Zolin[1], I V Burkova[2] and V V Zimin[1]

[1]Siberian State Industrial University, 42 Kirova Str., Novokuznetsk, 654007, Russia
[2]Institut of Management Problems n.a. V.A. Trapeznikov RAS, 65 Profsoyuznaya Str., Moscow, 117997, Russia

E-mail: zimin.0169@yandex.ru

Abstract. Models and mechanisms for project portfolio management, constantly generated by the IT service provider and designed to optimize IT processes of IT service life cycle stages are the subject of this article. The service improvement plan is formed from the project portfolio and is aimed at achieving critical indicators of IT services market success relevant for the current period. Two linear and one nonlinear models of the problem of forming a service improvement plan were considered. The criteria chosen for task optimization are absolute and relative value of the complex IT process efficiency indicator. Problems take into account the investment costs on implementing projects and changing the operational costs on functioning of optimized IT processes. All the problems were solved with the use of network programming methods.

1. Introduction

The structure of IT service lifecycle proposed in the ITIL-3 includes a continuous improvement stage. It is intended for improvement (in terms of efficiency and cost) the IT processes of the main stages (strategy, design, implementation and operation), which perform production functions, as well the support for service application by users.

Improvement projects are initiated both by the personnel of the stage, to whom this function had been delegated, and by specialists of the main stages. Each improvement proposal goes through several processing stages (filters). In particular, changes in the performance indicators of the improved IT processes are evaluated, and the impact of these improvements on changes in the performance indicators of the processes associated with improved process is assessed. Changes of operating costs (costs on functioning of the improved IT process) and the related processes are estimated. Note that not every project changes operating costs. A decision is ultimately taken on implementing or rejection the project proposal. All the agreed proposals enter into the portfolio of service improvement projects. Taking into account the allocated investments for the next planning period, the group for improvement services plans implementation of the most effective projects (forms the current service improvement plan).

2. Formalization of a problem on generation of a service improvement plan

Let $P = \{p_i \mid i = \overline{1,m}\}$ and $\{z(p_i), i = \overline{1,m}\}$ - are respectively the sets of projects initiated by personnel of different stages of IT service life cycle, and costs required for implementation of these projects.

The effectiveness $q_i = q(p_i)$ of an individual project, with regard to its contribution to the life cycle efficiency in the large, is determined by the values $\Delta\mu_{kl}^{n}(p_i)$ of changes in the performance indicators of the stages planned to be achieved, as a result of the project implementation.

In this case $n = \overline{1, N_{kl}}, l = \overline{1, L_k}, k = \overline{1, K}$, where N_{kl} - is a number of effectiveness indicators describing process 1 of the stage k, L_k - a number of process on the k-th stage, K –a number of life cycle stages. Having set weight coefficients $\alpha_{kl}^{n}(p_i), \sum_{n=1}^{N_{kl}} \alpha_{kl}^{n} = 1$ for effectiveness indicators of IT processes of all stages, weight coefficients $\gamma_k(p_i), \sum_{k=1}^{K} \gamma_k(p_i) = 1$ for the life cycle, and having made transformation of the local indicator measurement scales $\mu_{kl}^{n}(p_i)$ to a single discrete point scale of a given rank R, it's necessary to define effectiveness $q_i = q(p_i)$ of the individual project p_i for the life cycle in the following way [2]:

(1) $$q(p_i) = \sum_{k=1}^{K} \gamma_k(p_i) \sum_{l=1}^{L_k} \beta_{kl}(p_i) \sum_{n=1}^{N_{kl}} \alpha_{kl}^{n}(p_i) \Delta\mu_{kl}^{n}(p_i).$$

By Δz_{ikl} denote the change of operating costs on functioning of the process 1 of the stage k linked to the implantation of project pi. Subsequently, the total change Δz_i of operating costs linked to implantation of the i-th project will be:

(2) $$\Delta z_i = \sum_{k=1}^{K} \sum_{l=1}^{L_k} \Delta z_{ikl}.$$

Let $\Delta z^* > 0$ - is a change of operating budget admissible for the IT service provider in the next planning period. Let also z^*- are investments which may be directed to the implementation of optimization projects in this period.

Let us introduce a variable to formalize the problem at hand:

(3) $$x_i = \begin{cases} 1, \text{ if a project is entered in portfolio of accomplishable projects} \\ 0, \text{ if this is not the case} \end{cases}$$

3. Linear models of a problem for forming service improvement plan

The simplest model of the problem under study is the following (linear model 1):

(4) $$q(x) = \sum_{i=1}^{m} q(p_i) x_i \to \max$$

(5) $$z(x) = \sum_{i=1}^{m} z(p_i) x_i \le z^*.$$

It does not take into account changes in operating costs. Problem consist in definition of such $x_i, i = \overline{1,m},$

which maximize the criterion (4), but also meets the restrictions (5) on the total amount of investments. Task belongs to a class of knapsack problems and is solved by the dichotomy programming method [3,4].

Another model of the problem in question increasingly reflecting realities will be the following (linear model 2):

(6) $$q(x) = \sum_{i=1}^{m} q(p_i)x_i \to \max$$

(7) $$z(x) = \sum_{i=1}^{m} z(p_i)x_i \le z^*$$

(8) $$\Delta z(x) = \sum_{i=1}^{m} \Delta z_i x_i \le \Delta z^* .$$

Problem consist in determination $x_i, i=\overline{1,m}$, maximizing the criterion (6), but also meeting the restrictions (7) and (8), respectively, on the total amount of investments and on the admissible change of the operating budget in the planned period. The presence of two constraints does not allow the direct application of the dichotomy programming method to solve the problem (6) - (8). To make this possible, it is necessary to divide a plurality of projects $P = \{p_i \mid i=\overline{1,m}\}$ into two subsets:

(9) $$\{p_i|i=\overline{1,m}\}=\{p_i|i=\overline{1,m^o}\}\cup\{p_i|i=\overline{1,m^н}\}, m=m^o+m^н,$$

where m^o and $m^н$ – number of projects, which change and do not change operating expenses of processes, respectively.

Then the solution of task (6) - (8) can be reduced to the successive solution of three following problems:

1. Construction of dependency $\{q^o(x), \Delta z^o(x)\}$ of efficiency and operational cost changes for a subset $\{p_i \mid i=\overline{1,m^o}\}$. This requires to solve the task:

(10) $$q^o(x) = \sum_{i=1}^{m^o} q(p_i)x_i \to \max$$

(11) $$\Delta z^o(x) = \sum_{i=1}^{m^o} \Delta z_i x_i \le \Delta z^* .$$

Problem (10) – (11) – is a typical knapsack problem. Let $\{x^o\} = \{(x_1 x_2 x_3 x_4 x_5)\}$ - is a set of decisions of the problem (10) – (11), and $\{(q^o(x^o), \Delta z^o(x^o))\}$ - is a sets of related pairs of both a criterion value (10) and constraint (11).

2. Construction of dependency $\{q^н(x), z^н(x)\}$ of efficiency and operational cost changes for a subset of projects that do not change operational costs. This requires to solve the task:

(12) $$q^н(x) = \sum_{i=1}^{m^н} q(p_i)x_i \to \max$$

(13) $$z^н(x) = \sum_{i=1}^{m^н} z(p_i)x_i \le z^* .$$

It's the same typical knapsack problem. Let $\{x^н\} = \{(x_6 x_7 x_8 x_9)\}$ - is a set of decisions of the problem (12) – (13), and $\{(q^н(x^н), z^н(x^н))\}$ - are sets of related pairs of the criterion value (12) and constraint (13).

3. Formation of the set $\{x\} = \{(x_1 x_2 x_3 .... x_8 x_9)\}$, as a product of solution sets $\{x^o\} = \{(x_1 x_2 x_3 x_4 x_5)\}$ and $\{x^н\} = \{(x_6 x_7 x_8 x_9)\}$ of solutions, respectively, of problems (10) – (11) and (12) – (13):

(14) $$\{x\} = \{x^o\} x \{x^н\} .$$

4. Search on the set (14) of the optimal solution of a problem:

(15) $$q(x) = (q^o(x^o) + q^н(x^н)) \to \max$$

(16) $$z(x) = z^o(x^o) + z^н(x^н) \le z^*_{max}.$$

4. Nonlinear model of a problem

The q/z value (specific efficiency) is chosen often as an indicator of project efficiency, which shows what effect is per unit of cost. For that reason, it may be necessary to formalize the problem considered in the form of the following nonlinear model:

(17) $$q(x) = \sum_{i=1}^{m} q(p_i)x_i(\sum_{i=1}^{m}(z(p_i) + \Delta z_i)x_i)^{-1} \to \max$$

(18) $$z^*_{min} \le z(x) = \sum_{i=1}^{m} z(p_i)x_i \le z^*_{max}$$

(19) $$\Delta z(x) = \sum_{i=1}^{m} \Delta z_i x_i \le \Delta z^*.$$

The introduction of the left part of the restriction (18) is necessary, as in its absence the criterion (17) can choose the solution with unacceptably small use of the investment budget as the optimal solution.

The solution of the problem (17) - (18), in a similar manner to the method of solving the linear problem (6) - (8), is possible on the basis of decomposition into three subtasks:

1. Construction of dependency $\{q^o(x), \Delta z^o(x)\}$ of economical efficiency and operational cost changes for the subset $\{p_i \mid i = \overline{1, m^o}\}$. This requires to solve the task:

(20) $$q^o(x^o) = \sum_{i=1}^{m^o} q(p_i)x_i(\sum_{i=1}^{m^o}(z(p_i) + \Delta z_i)x_i)^{-1} \to \max$$

(21) $$\Delta z^o(x^o) = \sum_{i=1}^{m^o} \Delta z_i x_i \le \Delta z^*.$$

Let $\{x^o\} = \{(x_1x_2x_3x_4x_5)\}$ is a set of solution of problem (20) – (21), and $\{(q^o(x^o), \Delta z^o(x^o))\}$ is a set of related pairs of criterion values (20) and of restriction (21).

2. Construction of dependency $\{q^н(x), z^н(x)\}$ of economical efficiency and operational cost changes for a subset of projects that do not change operational cost. This requires to solve the task:

(22) $$q^н(x^н) = \sum_{i=1}^{m^н} q(p_i)x_i(\sum_{i=1}^{m^н}(z(p_i) + \Delta z_i)x_i)^{-1} \to \max$$

(23) $$z^н(x^н) = \sum_{i=1}^{m^н} z(p_i)x_i \le z^*.$$

Let $\{x^н\} = \{(x_6x_7x_8x_9)\}$ - is a set of solution of problem (22) – (23), and $\{(q^н(x^н), z^н(x^н))\}$ are sets of related pairs of criterion values (22) and of restriction (23).

3. Formation of the set $\{x\} = \{(x_1x_2x_3....x_8x_9)\}$, as a product of solution sets $\{x^o\} = \{(x_1x_2x_3x_4x_5)\}$ and $\{x^н\} = \{(x_6x_7x_8x_9)\}$ of solution, respectively, of problems (20) – (21) and (22) – (23):

(24) $$\{x\} = \{x^o\}x\{x^н\}.$$

4. Search on the set (24) of the optimal solution of a problem:

(25) $$q(x) = (q^o(x^o) + q^н(x^н))(z^o(x^o) + \Delta z^o(x^o) + z^н(x^н))^{-1} \to \max$$

(26) $$z^*_{min} \le z(x) = z^o(x^o) + z^н(x^н) \le z^*_{max}.$$

IOP Conf. Series: Materials Science and Engineering **865** (2020) 012016 doi:10.1088/1757-899X/865/1/012016

5. Example of solving linear and nonlinear problems

Tables 1 - 5 provide, respectively, initial data on IT process parameters of IT service life cycle and initial data on investment projects. This data are used to illustrate the procedures for solving linear problems (4) - (5) and (6) - (8) and nonlinear problems (17) - (19).

Table 1. Initial data on IT process parameter values, stage 1.

$\gamma_1 = 0,4$								
β_{1l}	$\beta_{11} = 0,4$			$\beta_{12} = 0,3$		$\beta_{13} = 0,3$		
α_{1l}^{n}	α_{11}^{1}=0,5	α_{11}^{2}=0,2	α_{11}^{3}=0,3	α_{12}^{1}=0,6	α_{12}^{2}=0,4	α_{13}^{1}=0,4	α_{13}^{2}=0,5	α_{13}^{3}=0,1

Table 2. Initial data on IT process parameter values, stage 2.

$\gamma_2 = 0,3$					
β_{21}	$\beta_{21} = 0,4$			$\beta_{22} = 0,6$	
α_{21}^{n}	α_{21}^{1}=0.8	α_{21}^{2}=0,2	α_{21}^{3}=0.2	α_{22}^{1}=0,5	α_{22}^{2}=0.3

Table 3. Initial data on IT process parameter values, stage 3.

$\gamma_3 = 0,2$							
β_{31}	$\beta_{31} = 0,3$			$\beta_{32} = 0,5$		$\beta_{33} = 0,2$	
α_{31}^{n}	α_{31}^{1}=0.6	α_{31}^{2}=0.2	α_{31}^{3}=0.2	α_{32}^{1}=0.4	α_{32}^{2}=0.6	α_{33}^{1}=0.9	α_{33}^{2}=0.1

Table 4. Initial data on IT process parameter values, stage 4.

$\gamma_4 = 0,1$							
β_{41}	$\beta_{41} = 0,4$			$\beta_{42} = 0,2$		$\beta_{43} = 0,2$	
α_{41}^{n}	α_{41}^{1}=0.2	α_{41}^{2}=0.5	α_{41}^{3}=0.3	α_{42}^{1}=0.7	α_{42}^{2}=0.3	α_{43}^{1}=0.2	α_{43}^{2}=0.8

Table 5. Initial data on portfolio of service improvement projects.

p_i	p_1	p_2	p_3	p_4	p_5	p_6	p_7	p_8	p_9
$q(p_i)$	0.592	0.384	0.420	0.212	0.15	0.576	0.816	1.374	1.334
$\Delta z(p_i)$	5	2	-1	3	-2	0	0	0	0
$z(p_i)$	14	16	19	9	8	10	13	21	7

According to the initial data, the first five projects change the operating costs on IT processes functioning, and the last four - do not change ($m^{o} = 5, m^{n} = 4$). Let us put $z_{max}^{*} = 90, \Delta z^{*} = 5$.

5.1 Results of solving linear problem 1

Having solved the problem (4) - (5) by the method of dichotomic programming, following three best results will be obtain, table 6.

Table 6. Optimal solution of the problem (4) – (5).

$x = x_1x_2x_3....x_8x_9$	1 1 0 1 0 1 1 1 1	1 1 0 0 1 1 1 1 1	1 0 1 0 0 1 1 1 1
q(x)	5.288	5.226	5.112
z(x)	90	89	84

5.2 Results of solving linear problem 2

Having solved the problems (10) –(11), (12) –(13) and (15) – (16) in accordance with decomposition schema , following three best results will be obtain, table 7.

Table 7. Optimal solution of the problem (6) – (8).

$x = x_1x_2x_3....x_8x_9$	1 0 1 0 0 1 1 1 1	0 1 1 0 0 1 1 1 1	0 1 0 1 1 1 1 1 1
q(x)	5.112	4.904	4.846
Δz(x)	4	1	3
z(x)	84	86	84

5.3 Results of non-linear problem solving

Let $z^*_{min} = 80$ .Having solved the problems (20) – (21) in accordance with decomposition schema for the first five projects $m^o = 5$, following results will be obtain, table 8.

Table 8 (a). Dependency $\{q^o(x), \Delta z^o(x)\}$ of economical efficiency and operational cost changes.

$x_1x_2x_3x_4x_5$	10101	10100	10001	10000	01111	01110
q^o	0.027	0.027	0.030	0.031	0.022	0.021
Δz^o	2	4	3	5	2	4

Table 8 (b). Dependency $\{q^o(x), \Delta z^o(x)\}$ of economical efficiency and operational cost changes.

$x_1x_2x_3x_4x_5$	01101	01100	01011	01010	01001	01000
q^o	0.023	0.022	0.021	0.020	0.022	0.021
Δz^o	-1	1	3	5	0	2

Having solved the problem (22) – (23) for following four projects, following results will be obtain, table 9.

Table 9 (a). Dependency $\{q^и(x), z^и(x)\}$ of economical efficiency and investment costs changes.

$x_6x_7x_8x_9$	1111	1110	1101	1100	1011	1010
$q^и$	0.080	0.063	0.091	0.061	0.086	0.063
$z^и$	51	44	30	23	38	31

Table 9 (b). Dependency $\{q^и(x), z^и(x)\}$ of economical efficiency and investment costs changes.

$x_6x_7x_8x_9$	1001	1000	0111	0110	0101	0100
$q^и$	0.112	0.058	0.086	0.064	0.108	0.063
$z^и$	17	10	41	34	20	13

Form a set (24) and determine on this set optimal solutions of the problem (25) - (26), table 10.

Table 10. Optimal solution of problem (25) – (26).

$x_1x_2x_3x_4x_5$ $x_6x_7x_8x_9$	1 0 1 0 0 1 1 1 1	0 1 0 1 1 1 1 1 1	0 1 1 0 0 1 1 1 1	1 0 1 0 1 0 1 1 1
q	0.058	0.056	0.056	0.056

z	84	84	86	82
Δz	4	3	1	2

All four solutions differ slightly in the q and z values. The choice of the best of them should be provided to decision maker. Let us notice that at the given basic data the best solution (1 0 1 0 0 1 1 1 1) of the nonlinear task (17) – (19) coincides with the best solution of the second linear task (6) – (8) and is the third in efficiency for the first linear task (4) – (5).

Note. Removing constraint on z^*_{min} for $z(x)$ will result in the selection of the optimal solution given in table 11.

Table 11. Optimal problem solution (25) - (26) in absence of constraint z^*_{min}.

$x_1x_2x_3x_4x_5$	1 0 0 0 0
$x_6x_7x_8x_9$	1 0 0 1
q	0.070
z	31
Δz	5

This decision, despite the significantly better value of the criterion, is unacceptable, as it uses only 30% of the investment budget.

6. Conclusion

Models and mechanisms proposed in this article for managing service improvements can be used by IT providers of all types as to improve existing IT processes, so to effectively adapt IT activities to the market environment.

References

[1] Zymin V V, Ivushkin A A et al 2013 *Basics of Life Cycle Management of Informatics and Automation Systems Services (Best Practices ITIL): A Tutorial for Universities* (Kemerovo: Kuzbasswuzdat) p 500

[2] Burkov V N, Condratiev V D and Schepkin A V 1912 *Mechanisms of Road Traffic Safety Improvement: Monograph* (Moscow: LIBRICOM) p 208

[3] Burkova I V 2012 *Method of Network Programming in Project Management Tasks* abstract of dssertation for degree of Doctor of Technical Sciences: 05.13.10 (Moscow: IPU) p 208

[4] Zymin V V, Kulakov C M and Zymin A V 2015 *Proc. of the X All-Russian Scientific and Practical Conference in Automation Systems in Education Science and Production* (Novokuznetsk: SibSIU) pp 228–33

IOP Conf. Series: Materials Science and Engineering **865** (2020) 012017 doi:10.1088/1757-899X/865/1/012017

Infrastructure management of software systems of modern "IT" in the automation of production processes

O N Nedzelskay, V V Moiseyev, A N Bikineeva and E N Bulakina

Siberian Federal University, Svobodny pr., 79. Krasnoyarsk, 660041, Russia

E-mail: elenagb09@mail.ru

Abstract. The article examines the methodology, describing the concept of information infrastructure management companies focused on their need, a comparison of models of information systems. The authors investigate the possibility of their joint use for maximum effect from the introduction of software systems. Process management of IT infrastructure of the enterprise has a number of advantages. Thus, in determining the types of activities, inputs, outputs, standards and requirements, an increase in the efficiency, the rationality of work can be achieved and quality management tools can be achieved. There are different methodologies for describing the achievement of business goals instruments with a high level of reliability of the information system. The most common was the British Library IT Infrastructure Library.

1. Introduction

To date, the majority of enterprises in our country use the standards for the organization of the information departments. It should be noted that the use of international practices and methodologies cannot run counter to the standards that must be met to the enterprise. Comparison of different approaches to the organization structure of enterprise information quality standards require changing view of IT, namely IT services consideration not as an opportunity for service, but as a unit of the enterprise using IT assets. Thus, the purpose of IT services will be to ensure the work of IT assets, repairs, maintenance work, elimination of emergency situations. Indeed, the Service Desk modules are most commonly used and the configuration database for the accounting technique of ITIL system. It should be noted, that in this case, these modules are used for the organization of work according to ISO 18322, which focuses on the elimination of failures and accounting techniques. Approach, focusing on maintenance and repair of IT assets, it is used in most enterprises, state and municipal institutions.

There are different methodologies for describing the tools to achieve business goals with a high level of reliability of information systems. Information technology, to date, is developing rapidly. This applies both to improving element basis, as a result, increasing computing power; to improve the system and application software, the implementation of mainframe technology in desktop personal computers (e.g., multi-threading, virtualization); to expanding the global Internet, which is available all the greater part of the world's population. Development of information technology provides growing opportunities for business relationships. The use of advanced technology gives a competitive advantage to the enterprise only if the most effective implementation of these technologies. Results of operations for the automation of production processes to meet the requirements put in the beginning of automation of the enterprise is: increasing productivity, optimizing management processes, improve product quality, to meet the new requirements. Automation of the process may have varying degrees of maturity: from chaotic to carried out in accordance with the best practices of global companies. The effect of this process on the quality of the end result is undeniable. In a qualitative implementation of software system that meets the requirements of the enterprise, as a result will be achieved the desired implementation of quality indicators. When chaotic introduction of even the best software package available mass of bugs admitted at all stages of implementation.

2. System approach to automation of software systems in the workplace

There are several approaches evaluation software product quality indicators, such as the evaluation of internal and external quality indicators mixed approach [1]. Indeed, it is important when choosing automation guide program analysis tools. But the most important factor in achieving positive results in the automation activity is a systematic approach to all work as a whole.

The most common was the British Library IT Infrastructure Library (ITIL), COBIT (Control Objectives for Information and Related Technology ("Objectives for Information and Related Technology"), a methodological model of Microsoft Operation Framework (MOF). To date, the majority of enterprises in our country use standards for the organization of the information department. There are a number of standards to be applied to equipment maintenance, such as ISO 18322 (in Russian GOST 18322) standard management ISO 9004-2010 quality, the creation of GOST 34.601-90 automated systems to processes of life Ikla software ISO / IEC 12207-2010 and other [2-4].

Compliance with standards is considered to be one of the conditions necessary for the confirmation of the high quality of produced goods or services. In some ways, a certain conformity to strict standards makes it difficult, and sometimes even eliminates the opportunities of development, previously considered impossible to achieve performance metrics. One prominent example is the achievement of international fame due to high quality company Toyota as a result of the practice of Kaizen, also known as "lean production". Today, many global companies have the experience of "lean production" and, thus, achieved outstanding results. It should be noted that over the past 10 years, many domestic companies have implemented or have taken the first steps towards the implementation of "lean production" and have been able to assess the results.

It should be noted that the use of international practices and methodologies cannot run counter to the standards that must be met to the enterprise.

Comparison of different hikes to the organization structure of enterprise information quality standards require changing view of IT, namely IT services consideration not as an opportunity for service, but as a unit of the enterprise using IT assets. Thus, the purpose of IT services will be to ensure the work of IT assets, repairs, maintenance work, eliminate emergency situations. Indeed, the Service Desk modules are most commonly used and the configuration database for the accounting technique of ITIL system. It should be noted that in this case, these modules are used for the organization of work according to ISO 18322, which focuses on the elimination of failures and accounting techniques [5]. Approach, focusing on maintenance and repair of IT assets, is used in most enterprises, state and municipal institutions.

As far as ITIL and MOF approaches, these approaches are fundamentally different from those used to stay with us, these approaches are similar, in some ways have serious differences. ITIL uses five elements lifecycle service strategy 1-, 2-projecting service 3-introduction service, service operation-4, 5-monitoring service. Continuous improvement processes satisfying PDCA model (Plan-Do-Check-Act). The core of MOF consists of three phases: planning, implementation, operation, which complement the level of control that is used for all life cycle phases [6].

It should be noted that ITIL is designed so that it can be used regardless of the case used in the technology or system. As a result, ITIL focuses on the management structure, allowing IT to work successfully in the enterprise, regardless of the technologies used. In turn, the MOF is built for success on the Microsoft platform. It believed that MOF can be applied for use with other platforms [7].

ITIL technology uses a process approach, most of the components describing the processes and functions. Processes characterized measurability, certain result, the orientation of the consumer, the trigger (figure 1).

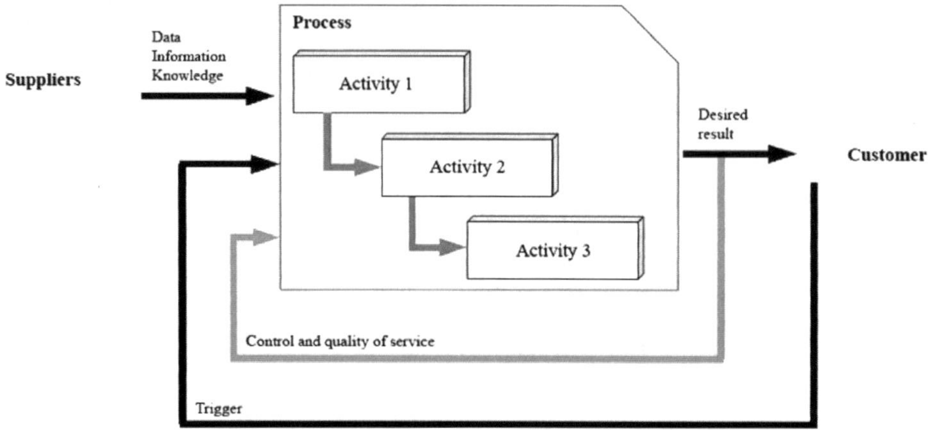

Figure 1. Description of the process in ITIL.

Process management of IT infrastructure of the enterprise has a number of advantages. Thus, in determining the types of activities, inputs, outputs, standards and requirements, you can increase the efficiency, the rationality of work, getting quality management tools.

3. The process of managing IT infrastructure - as the quality management process

Structuring process allows to control the quality in accordance with the cycle of the Deming PDCA (figure 2).

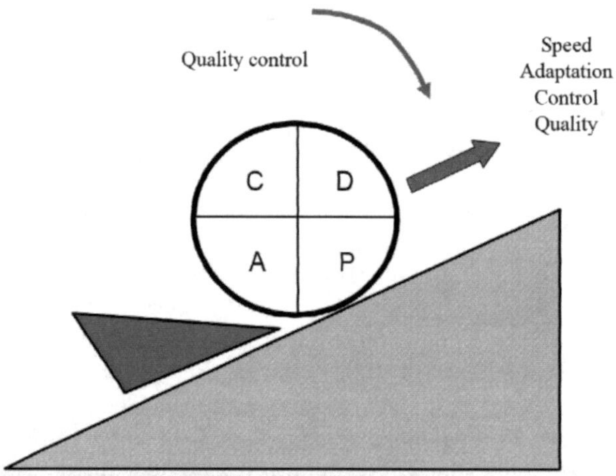

Figure 2. Deming cycle.

Thus, ITIL MOF and methodology can be used for process control and quality control of the information processes for the enterprise.

But how the organization will manage its risks and ensure the security of information resources on which it is dependent? As an organization can be confident that the information structure is achieving its objectives and supports the development of the company? These questions are answered COBIT – a package of open documents, more than 40 international and national standards and guidelines in the field of IT management, audit and IT security.

In difficult conditions, the successful and rapid decision-making in respect of assets and risks of the enterprise is necessary to choose what and how should be measured. Enterprises need objective criteria for assessing their current state, and the improvements they require, as well as some tools with which you can make improvements. To answer these questions, COBIT provides the following definitions:

All other paragraphs are indented ("DOS style text is indented"):

- Maturity model, which is expressed in the form of a comparative analysis of the efficiency and capacity of IT processes.
- Targets and indicators of IT processes necessary to identify and assess their results and efficiency, based on the principles of a balanced system of business indicators, proposed by Robert Kaplan and David Norton.
- Action Objectives for direct IT process management, based on the objectives of COBIT control [8].

Evaluating the effectiveness of the IT sector is part of the control (figure 4). Evaluating the effectiveness considered COBIT and includes setting and monitoring measurable objectives which define the processes of IT and capacity and process efficiency. In many studies it has been found that the defects in the transparency IT costs, risks, and determining values are among the main incentive to improve the management area of IT, figure 3. While other management are subject to evaluate the effectiveness, transparency is achieved primarily with her help.

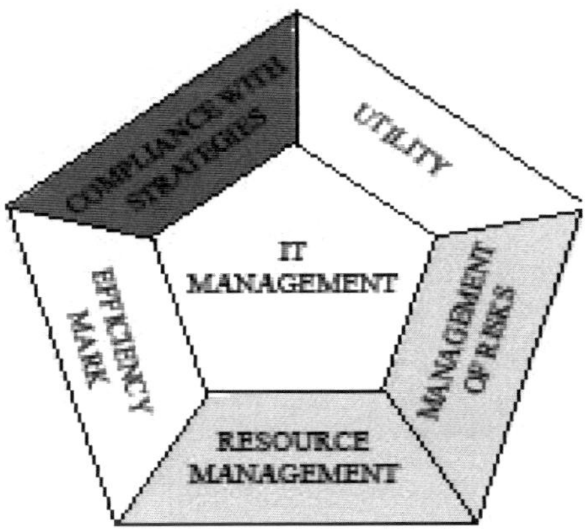

Figure 3. Key areas of IT management.

- Compliance with business strategy is a priority area. This area is responsible for communication between the enterprise goals and IT enterprise.
- Risk management and resource management are in second place. The key issues of resource management is the optimization of infrastructure, investment, knowledge, library staff. Risk management provides leadership, the company information on the possible risks related to the company.
- Evaluating the effectiveness provides control over the implementation of the strategy, the project results, the use of resources, processes, efficiency and service. For this purpose, in particular, a balanced scorecard that convert strategy in sequence, the results of which are measured by other, compared with accounting, methods.
- Usefulness provides control over the IT to processes provide advantages, certain strategy reduced costs.

4. Speed, adaptation, management, quality. Integrating the use of IT services

These areas of enterprise IT management tasks describe works with senior management, implementing IT governance in their organizations. COBIT methodology provides a common process model that represents all the processes as elements of IT functions that makes this basic model understandable to operational IT staff and management. This process model is correlated with the focus areas of IT governance responsibilities and connects operations personnel with the guidance [9,10].

The COBIT focuses on what is required to achieve adequate management of IT at a high level, it is linked to other, more detailed standards in the IT sector, as well as best practices and methodologies.

COBIT methodology acts as an integrator of guidance materials, summarizing key objectives under a single methodology, which, in turn, linked to the management and enterprise requirements (figure 4).

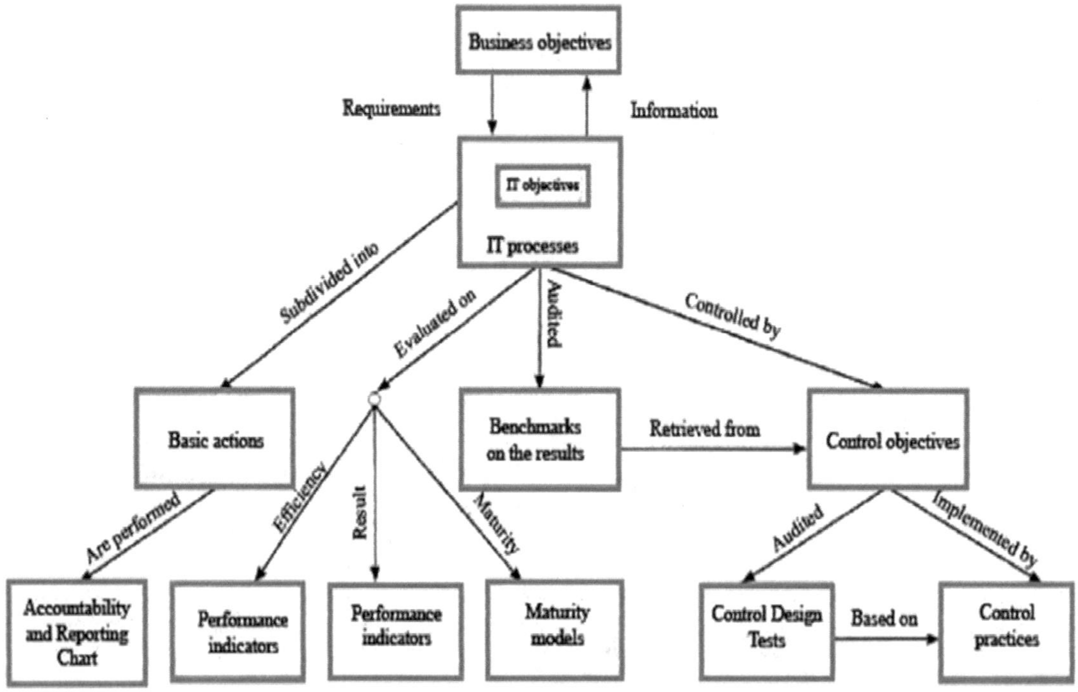

Figure 4. Scheme relationship COBIT components.

The COBIT focuses on what is required to achieve adequate management of IT at a high COBIT is a methodology and tool that allows managers to address the shortcomings with regard to the requirements of control, technical issues and risks, and to convey the current level of control to stakeholders enables develop clear policies and best practices for IT control in organizations.

Table 1 shows the matching problems and the methodologies used.

Table 1. Compliance processes and methodologies.

Level	Appropriate methodology / standard
Audit of the management and control system	ISO 20000: the process of comparing the test results with the criteria
management control	COBIT: management and process management control
Process Management Service Management	ITSM: process management (ITIL, MOF, and others)
enabling technologies for the management of the customer (services)	ITSM: Quality management services
infrastructure management	Technical competence

The company may pursue different goals when implementing new methodologies: the creation of a high-speed system, reducing risk, reducing costs, audit and others. How to evaluate the effectiveness of the implementation methodology? It depends on the task. For example, you can use Key

Performance Indicators - KPI, IT-related processes. To do this, we can pick up the KPI indicators that are the most related to the field of automation. For example, the introduction of call-center should be reduced response times to inquiries, registration, and other applications. After defining the KPI's, quantitatively related to the IT services we can collect statistics, to track the dynamics of the specified parameters. Monitoring on the long stretches of time will allow to draw a conclusion on quantitative indicators, and to assess the possibilities for changing the test indicator. Such statistics will evaluate the efficiency of the implementation, as well as be used for further work on the automation. On its basis we will be able to make preliminary estimates predict the effectiveness of similar systems, make decisions about the appropriateness of process automation.

Formulas of economic efficiency indicators of the implementation of information management systems structure can be also used for the calculation:

$$P_{IT} = (T_0\text{-}T_S)_{qh}, \quad (1)$$

where P_{IT} – is an indicator of IT productivity, which shows how efficiently the time is used by employees of the IT department;

T_0 – the time that one IT employee spent on average "manually" performing work before implementing an IT management system;

T_S – the time that one IT employee should spend on average "manually" performing work after implementing an IT management system;

q_h – average hourly earnings of performers.

$$E_{IT}=(N_0\text{-}N_S)q_y, \quad (2)$$

where the E_IT – an indicator of the effectiveness of IT service personnel, which characterizes the reserve of personnel preparedness: what additional number of different technologies and devices an IT department can handle without an increase in the number of employees;

N_0 – the number of IT service employees needed to support additional devices, applications, and users before implementing an IT management system;

N_S – the expected number of IT service employees needed to support additional devices, applications, and users after the introduction of an IT management system;

q_y – the value of the annual salary of IT specialist.

Then increase user productivity while reducing downtime E is expressed by the formula:

$$E = (W_0\text{-}W_S)\ s_h, \quad (3)$$

where W_0 – the time during which users do not have access to the applications necessary for the performance of their duties, before the implementation of the IT management system;

W_S – the expected time during which users will not have access to the applications necessary for the performance of their duties, after the introduction of an IT management system;

s_h – the average hourly wage of employees of the enterprise.

In such a calculation the economic indicators are summarized in each direction. If the deduction of implementation and operation of the system of this expenditure, we obtain an estimate of the effectiveness of implementation.

5. Conclusion

Changes related to IT infrastructure with enterprise asset at the service allows you to make the IT department the most relevant requirements of the time, as a result, to ensure the effectiveness of investments in the sphere of IT investments. The use of various technologies for the optimal functioning of IT departments is a complex task. When choosing methodologies, one should start from the goals that must be achieved. The most complete and structured methodology in the field of service management and ITIL service management library can be used, focused on process management,

IOP Conf. Series: Materials Science and Engineering **865** (2020) 012017 doi:10.1088/1757-899X/865/1/012017

which solves the management of infrastructure control and management system, has a positive effect on customer satisfaction.

References

[1] Bulakina E N, Nedzelskay O N, Moiseev V V and Bikineeva A N et al 2017 *IOP Conf. Series: MSE* **411** 012050
[2] Azaryeva V V and Stepanov S A 2013 *"Quality Management": Proc. of the Regional Sci. and Pract. Conf.* (Voronez: Voronezh Gos. Tekhn. Un-t) pp 101–104
[3] Dukhanov A V 2010 *Simulation Modeling of Complex Systems* (Vladimir: Vladimir State Un-ty) p 115
[4] Jan Van Bong 2012 *IT Service Management, Introduction* (SMF)
[5] Jacob D Bekenstein 2013 *Scholarpeda* http://www.scholarpedia.org/article/Bekenstein_bound
[6] Jerry Wu and Yin-Lin Shen 2013 *Computational Intelligence and Soft Computing*
[7] A survey among 565 IT directors in the US and Europe 2007 Source Desktop and Application Virtualization, Forester Research Inc.
[8] Moiseev V V, Bulakina O N and Bulakina E N 2013 *Fundamental Research* 6 (part 3) 709–14
[9] Moiseev V V, Ketov A V and Lebedkin P V 2010 Proc of the VII All-Russian Conf. of Students, Graduate Students and Young Scientists in Microsoft Technologies in Programming Theory and Practice (Tomsk: TPU) pp 126–128
[10] Bulakina E N, Nedzelskay O N, Moiseev V V and Bikineeva A N et al 2019 *IOP Conf. Ser.: Earth Environ. Sci.* **377** 012014

Generalization of experience and classification of risks of IT services by the stages of the life cycle

T V Kiseleva and E V Maslova

Siberian State Industrial University, 42 Kirova Street, Novokuznetsk, 654007, Russia

E-mail: kis@siu.sibsiu.ru

Abstract. The article discusses several classifications of risks for various reasons. In addition, a classification of information risks by stages of the life cycle of an IT service is proposed.

1. Introduction

At present, the issue of ensuring information security is one of the most acute. As a result of the widespread use of information technologies at enterprises and organizations of various levels, there is a possibility of damage, loss, theft of information assets. To reduce the damage that may occur, risk management or risk management methods are used.

One of the stages of successful risk management is the identification of risks and the subsequent classification according to the selected criteria. First of all, it is necessary for the application of the most rational and optimal protective measures. But there is still no consensus on the identification of signs and principles of classification.

2. Results and discussion

According Batova I.V. [1] important for the classification of risks are the time of their occurrence, factors influencing this, the place where they arise, the scope, nature of the consequences, the size of possible losses. Table 1 provides a detailed overview of the features of the proposed Batova I.V. risk classification.

Table 1. Classification of risks.

Attribute	Types of Risks
Time of occurrence	Retrospective, current, promising
Occurrence factor	Political, Economic
Place of origin	External, internal
Sphere of origin	Entrepreneurial, investment, etc.
Hazard type	Man-made, natural, mixed
Occurrence rate	Macro, micro, mesoscale
Degree of certainty	Known, predictable, unpredictable
Stage of occurrence	Design, planned, actual
Degree of validity	Reasonable, unfounded
Size of possible losses	Acceptable, critical, catastrophic
Scale of consequences	Global, regional, local
Legal conditions of occurrence	Risks arising from obligations and arising for other reasons

In [2], it is proposed to classify risks according to two principles, within which risks are distributed into groups according to a certain criterion. The first classification is called substantive, since risks are classified in this case according to their specific content. In the subject classification, risks are divided into groups, followed by fragmentation into species, varieties (table 2).

This classification has disadvantages. Firstly, in this case it is practically impossible to single out all potential types of risks. In addition, when using this classification, duplication is inevitable, and it also does not provide guidance on which management methods to be guided in the first place.

Table 2. Subject classification of risks.

Feature	Species	Content
By the nature of the consequences	Clean	Risks that, if implemented, do not lead to a positive result
	Speculative	If risks are realized, a positive result is possible.
Due to occurrence	Natural	Risks that may be realized as a result of the action of natural processes and phenomena
	Environmental	Risks depending on the state of the environment and liability for environmental damage
	Political	Risks associated with the socio-political situation in the country
	Social	Risks of negative public events
	Liability risks	Risks associated with the occurrence of civil liability for damage
	Transport	Risks associated with the carriage of goods
	Information	Information Technology Risks
By area of origin	Commercial	Industrial, commercial, property, legal
	Financial	Investment, direct loss risks, purchasing power risks
	Innovative	Risks Associated with Uncertainty in Innovation

The second mentioned principle of risk classification in work is a breakdown of risks by source and stage of occurrence. Such a classification is often used when choosing a risk management strategy, therefore it is called managerial, an example of the described classification is given in table 3 [2].

Table 3. Management classification of risks.

Attribute	Kinds	A comment
By nature of occurrence	Subjective	Risks associated with the identity of the decision maker, his level of education, professionalism, errors
	Objective	Risks associated with changes in legislation, natural disasters, and other external circumstances
Depending on the stage of commercial activity	At the decision making stage	Risks associated with errors and inaccuracies in the source information
Depending on the stage of commercial activity	At the stage of implementation of the decision	Risks due to errors made during the implementation of the decision, or unexpected changes
By scale	Local (design)	Risks of an individual organization or a specific project
	Industry	Industry Specific Risks
	Regional	Risks at the level of constituent entities of the Russian Federation
	National	Macroeconomic Risks
	International	Risks associated with international relations, global natural disasters
By area of origin	External	Environmental Risks
	internal	Risks associated with the specialization of the organization

Possible insurance	Insured	Risks are quantifiable and accepted by insurance companies
	Uninsured	Risks not measurable, not accepted as a result of this by insurance organizations
If possible diversification (typical negative consequences)	Systematic	The risks inherent in this type of activity, which are regular, are based on various laws.
	Specific	Risks associated with a specific operation caused by rare circumstances
By degree of admissibility	Minimum	Risks associated with a small level of losses (up to 10% of profit)
	Elevated	Risks associated with losses not exceeding 25% of profit
	Critical	Risks associated with losses from 50% to 75% of profit
	Invalid	Risks associated with the loss of 75 - 100% profit
By hazard	Technogenic	Risks associated with human activities
	Natural	Natural disaster risks
	Mixed	Risks associated with natural events caused by human activities
As detailed as possible	Simple	Risks that cannot be broken down into small factors
	Complex	Risks that include a risk group
By time	Perpetual	Permanent Risks
	Urgent	Risks acting for a limited time or at a time

The paper [3] provides a classification of information risks, which, according to the author, are included in the group of investment risks. Information risks, in turn, are divided into business planning risks. The author of the work classifies these risks according to the following criteria:

1. Functional risks that are determined by one or another part of the IP of the enterprise, that is, the risks of collecting, processing, presenting information, etc .;
2. Structural risks are characterized by the structure of the enterprise, these include the risks of accounting or management accounting, planning, control, etc .;
3. Temporary risks determined by one or another stage of the project life cycle, that is, risks of development, coordination, implementation, operation;
4. The risks of influence, which are determined by the human factor and are divided into random and forced.

Figure 1 illustrates this classification.

The general classifications of all possible risks, which are given in the literature devoted to this issue, were considered above. The report examined the informational risks to which IT services are exposed. Based on the above information, as well as analysis of the reported data on incidents of IT incidents of a large metallurgical enterprise, a classification of possible risks was developed, divided by the stages of the IT service life cycle, which is shown in table 4 [4, 5]. The proposed classification allows us to more accurately indicate the causes and sources of risk realization, therefore, to choose more rational protective measures.

Classifications are also possible according to other grounds, for example: by the method of elimination, by the culprit, by the place of origin, by the degree of influence on the work of the enterprise, by the degree of complexity of recovery from exposure, by threats, by consequences.

It is important to remember that the correct classification of risks helps to avoid double-counting, which means unnecessary or unforeseen costs.

Table 4. Classification of risks by stages of the life cycle of IT service.

Stage	Name of risk	Stage Name of risk Comment (reasons)
Strategy	Risk associated with incorrect	Service Requirements Have Changed

	assessment of service demand	Incorrectly defined assets, goals, plans, budget
Design	Asset mispricing risk	Asset shortage
	Risk associated with untimely launch of an IT service	Unreliable equipment supplier
	Deadline risk	The timing and complexity of the work are incorrectly defined, therefore, the implementation period is delayed
	Risk of non-acceptance of IT service by the customer due to changes in customer service requirements	Changes in external conditions, changes in situations at the customer
Implementation	Technical risk	Software / hardware is out of date
	The risk of total or partial loss of IT service	Force majeure circumstances: fire, flood, earthquake, viral attacks
Exploitation	Long time lack of access to IT service	The recovery time after a failure is longer than expected
	Risks associated with violation of data integrity, confidentiality and availability	Unauthorized access
	Risk associated with non-fulfillment of agreements between the customer and the IT provider	Lack of support from the provider
	The risk of total or partial loss of IT service	Force majeure circumstances: fire, flood, earthquake, viral attacks

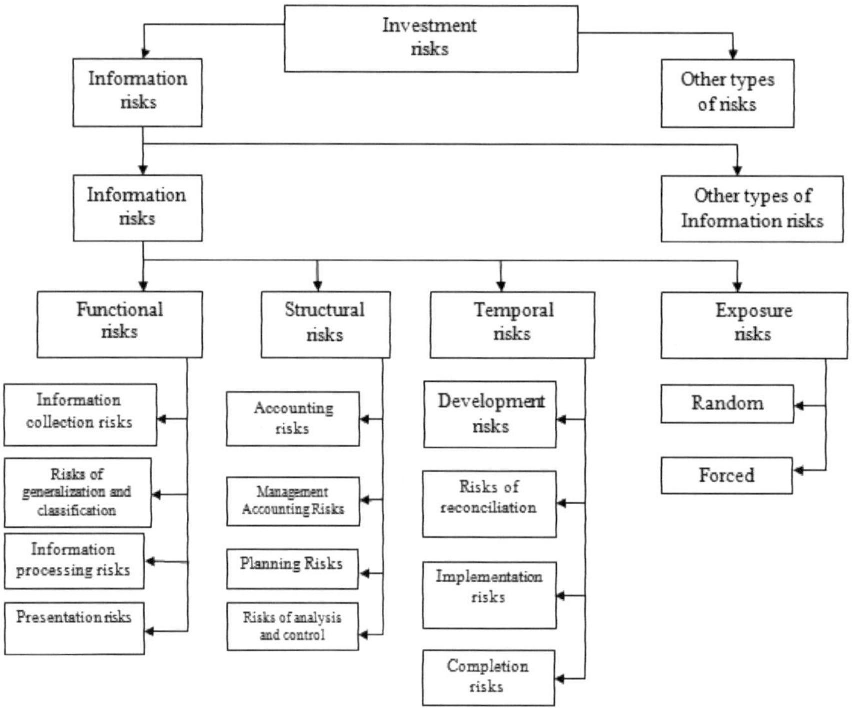

Figure 1. Classification of information risks.

3. Conclusion

The report provided an overview of the available risk classifications by various attributes. And also a classification of the risks of the IT service according to the stages of its life cycle is proposed. Using this classification, at the stage of developing protective measures, it is possible to choose those measures that are most optimal for specific types of risk and will minimize the possible damage from their implementation.

References

[1] Batova I V 2015 *International Student Scientific Bulletin* **1** 1–8
[2] Panyagina A E 2012 *Modern Economics: Problems, Trends, Prospects* **6** 1–11
[3] Hasanov G M 2009 *TDR* 3 151 – 153
[4] Kiseleva T V and Maslova E V 2015 *Proc. of the VIII Int. Conf. Modern Methods of Applied Mathematics, Control Theory and Computer Technology* (PMTUKT-2015) (Voronezh) pp 180–182
[5] Kiseleva T V and Maslova E V 2014 *Proc. of the Int. Sci. and Practical Conf. in Education, Society: Trends and Prospects* (Moscow) vol 3 pp 67–68

IOP Conf. Series: Materials Science and Engineering **865** (2020) 012019 doi:10.1088/1757-899X/865/1/012019

Automated visualization of calculation results in the Algozit programming environment

V O Kaledin, A E Paulzen and A D Ulyanov

NBI of Kemerovo State University, 23 Tsiolkovsky Street, Novokuznetsk, 654000, Russia

E-mail: zbrg@mail.ru

Abastract. The functional object diagram of algorithms for constructing a graphical representation of the source and calculated data in the visual programming environment "Algozit" is considered. When defining the functional content of circuit elements (algomats), a built-in interpreted programming language is used, supplemented by graphical operators. Using the built-in language simplifies the modification and support of the developed visual algorithm. Graphic operators use the capabilities of a graphic library built into the visual programming environment. The library allows you to accumulate data on graphic primitives in three-dimensional space and build them in a graphic window in the form of a central projection or stereo image. The graphics window is equipped with display controls.

1. Introduction

Visual programming environment "Algozit" [1, 2] allows you to design computational algorithms and general purpose algorithms. Elements of the functional object network (FON) are algomats. The functions of the algomats are defined in the form of methods of functional classes in C ++, which are defined in the library – "factory of objects". The interpreted language "Jadro" [3] was integrated into the "Algozit" environment to speed up the development process and make the execution of executed algorithms more clear. This allows you to make changes to the functional content of algomats and not to resort to changing compiled program code in C ++. This modification allows you to reduce the cost of working time of the developer.

In the study of material objects by numerical modeling methods [4], the problem arises of visualizing the initial and obtained data. Mathematical methods and hypotheses can have significant modifications. The format of the initial data of the model and calculation results may depend on these modifications. Often an individual solution is used to visualize data. To speed up the process of selecting or developing the necessary visualization algorithm, a universal FON was developed in the "Algozit" environment. To build images using the built-in language "Jadro" and its graphical tools.

2. Description of implementation

Figure 1 partially shows the communication structure according to the components of the "Algozit" environment: block 1 in the figure indicates the "Debugger" module with the FON loaded in it for execution; block 2 – library-factory of functional classes, in the methods of which either the compiled code is executed, or – the script code through the interpreter; block 3 – interpreter module of the embedded language "Jadro", which is equipped with graphical operators; block 4 – graphic library, including the functions of display on the screen, viewing settings and legends.

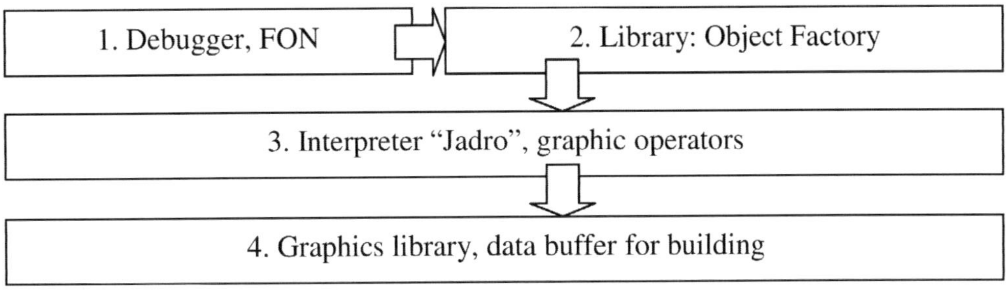

Figure 1. The dependency diagram of the modules of the functional-object programming environment "Algozit".

The developed FON can be considered universal due to general algorithms for the accumulation of topological and geometric data and the availability of modification of network parts for preparing portions of graphic data of various kinds. Algorithms are implemented using the "Jadro" language and its graphical tools in algorithms handlers. These algorithms are available for editing in the runtime constructor of the FON of the "Algozit" environment.

In order to implement support for the core language, the ability to replace existing C ++ handlers with scripts with predefined variables has been added to the settings of the algomats. An example of the script settings window is shown in figure 2. Each table item in this window corresponds to a specific method of the algomat. Predefined variables in scripts and their types are defined according to the selected functional class of the algomat.

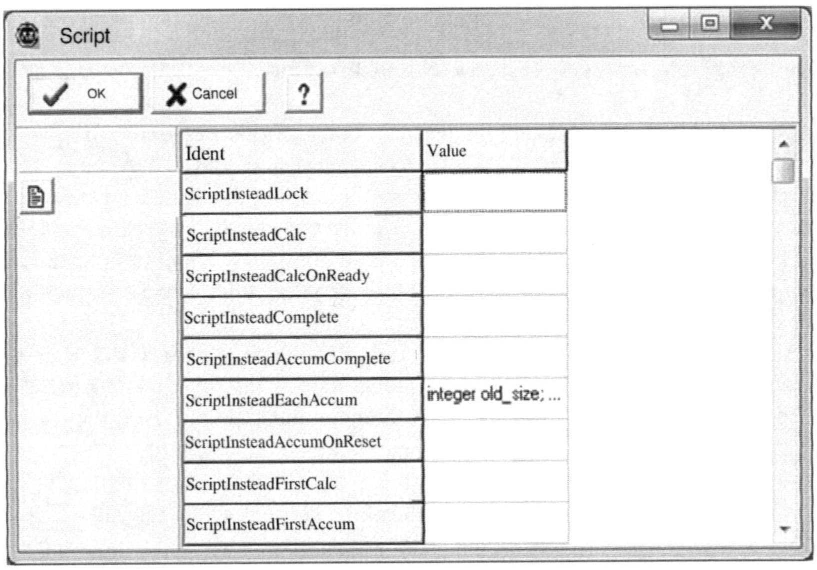

Figure 2. Algorithm script settings window.

The algorithm for accumulating data for drawing was implemented in the form of FON. Its upper level is shown in figure 3.The results of the calculation in a tabular form are loaded into the memory by the network panels, processed and accumulated in the memory of the algomats "Stress", "Movement", "Connection of circuits" and "Form". The names of the panels correspond to their content. In the figures, the names of the algomats are abbreviated. They serve as input data for image building. Data for construction are formed in the form of points, their attributes and lines in three-dimensional space by the following language operators: AddContour – add contour, AddNode – add point to contour, SetValueOnTitle – adding point attribute data. Buffer management by commands:

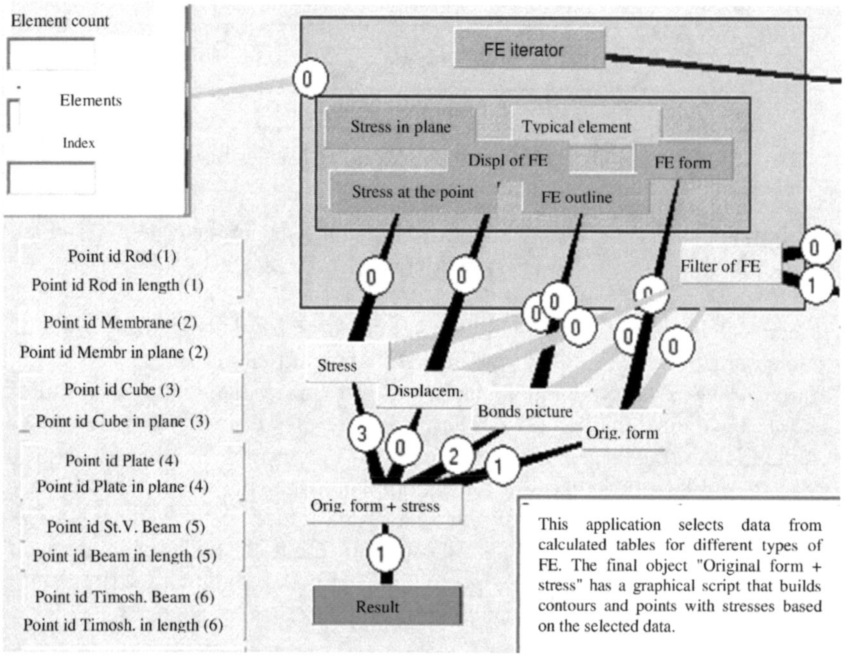

Figure 3. FON top-level for data accumulation and output to the graphics window.

ClearImage – clear image data, CompleteImage – completion of construction.

Before you start drawing, you must reset the previously accumulated graphic information with the ClearImage command. The AddContour command is executed to create a new sequence of points (contour). Its result is a new contour number. After the contour is created, it can be replenished with dots (AddNode). When executing the CompleteImage command, contours are drawn in the order of addition with all points with the specified color and drawing style. The SetValueOnTitle command is used to add point attribute data.

Attribute data is used to construct scalar or vector fields for the selected data component. The script for graphical construction using the prepared data is implemented in the “Script Instead of Drawing” handler of the “Original form + stress” object and is executed when the build command is called during the execution of the algorithm in the “Debugger” module [1,2].

Figure 4 shows the result of constructing the initial and deformed shapes with scaling of the linear components of displacements in the window of the Graphics module. The original shape is drawn with dashed lines in gray, the shape with solid movements taken into account is constructed with black lines. The components of the source or calculated fields at the internal points of the contours can be represented by colored dots with a legend.

The window contains interactive viewing controls: rotation of the object relative to the coordinate axes, setting the color legend for scalar fields, construction of vector fields and diagrams, building a deformed shape relative to the original, display image element captions, anaglyph stereo image.

The constructed scheme (FON) refers to the finished tabular data. Determining the method of drawing and selecting points from a piece of data is carried out separately for each view, by means of switchable FON pages.

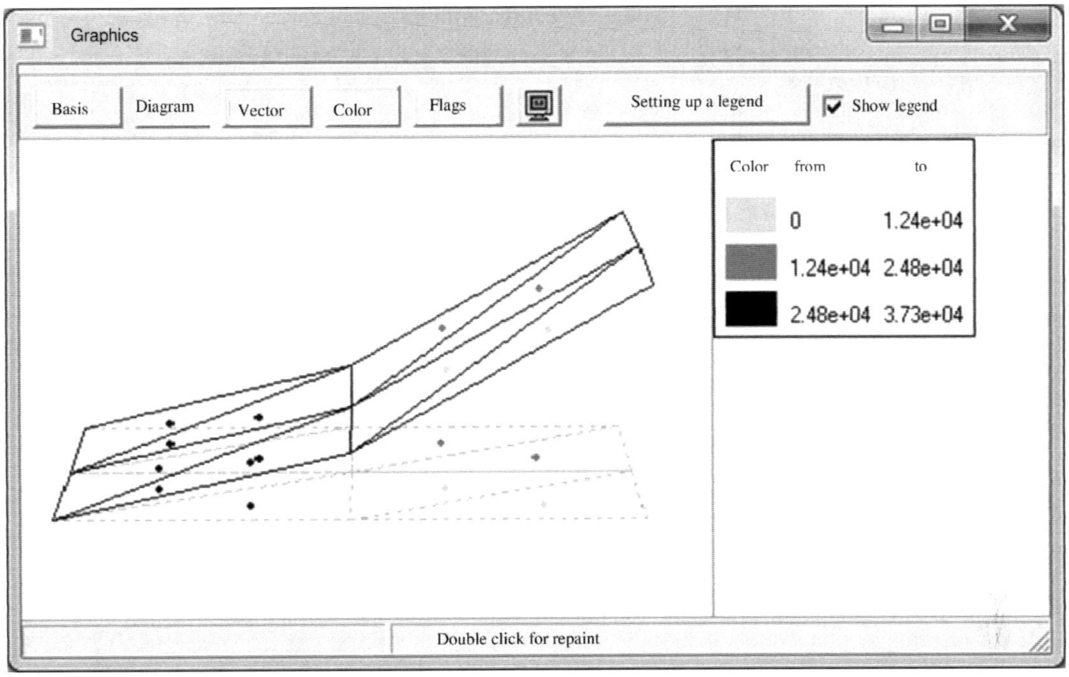

Figure 4. An example of a graphical construction of a model of a plate with stresses and displacements of the scale.

Figure 5 shows an example of one of the switchable web pages. Each switchable page (subnet) provides the correct selection of data from the same sources as in the calculations. Therefore, the network can be adapted for any algorithms implemented within the framework of the considered programming environment. The accumulation of data is carried out in the algorithms "Stress subsampling" (stress table at points of the current element) and "Stress subsampling plan" (stress table at the plan points of the current element). The accumulation algorithm is set by a script in the "Jadro" language.

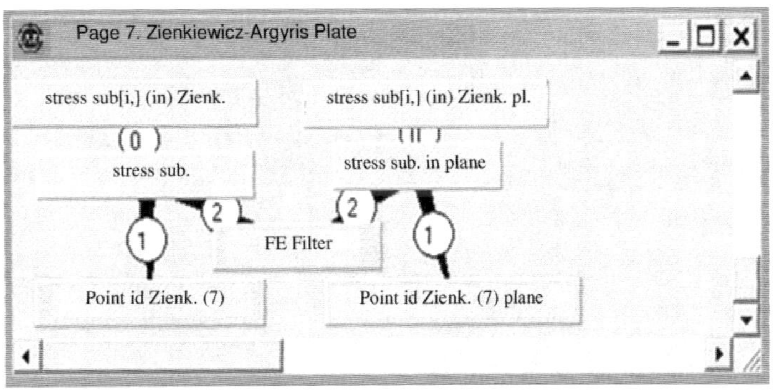

Figure 5. A fragment of a nested switch network that implements data accumulation for drawing.

3. Conclusion

Thus, the developed standard FON allows visualization of calculation results automatically for many data types. If the data type is not supported by the FON, then the user has the option to enable its support. To do this, you need to add a small number of typical algomats with functional content in the "Jadro" language to a switchable page.

Acknowledgments
This work was partially supported by the grant from the NBI of KemSU (grant agreement No. 3-05 / 1GR-18 dated 10/18/2018).

References

[1] Ulyanov AD, Kaledin V O et al 2016 *Proc. of XVI International Scientific and Practical Conference* in 3 parts (Chita: Transbaikal State University) part 2 pp 141–145

[2] Ulyanov A D and Kaledin V O 2016 *Proc. of the V Int. Sci. and Pract. Conf. in Prospects for the Innovative Development of the Coal Regions of Russia* (Prokopyevsk: Branch of KuzSTU) pp 350–352

[3] Kaledin V O *Interpreter "Jadro"* The Certificate on Official Registration of the Computer Program No 2017612706 publ. 02.03.2017

[4] Bathe K Ju 1996 *Finite Element Procedures* (New Jersey: Prentice Hall) p 1050

Toward industry 4.0 in energy sector

Elena Goosen[1], Sergey Nikitenko[2], Elena Kagan[1] and Evgeniya Pakhomova[2]

[1]Kemerovo State University, 6 Krasnaya street, Kemerovo, 650000, Russia
[2]Federal Research Center of Coal and Coal Chemistry, Siberian Branch of the Russian Academy of Sciences, 18 Sovetskiy Ave., Kemerovo, 650000, Russia

E-mail: nsm.nis@mail.ru

Abstract. The article attempts to identify the profound changes occurring in the global and Russian energy sectors under the conditions of Industry 4.0. The features of the implementation of the global innovative Industry 4.0 project in the global energy sector are studied; a change in the technological basis, an upgrade of value chains and a change in business models are shown. The state and prospects of the Russian fuel and energy complex transition to a new Industry 4.0 development paradigm are analyzed. The main models and barriers to the formation of new value chains in the Russian fuel and energy complex are identified. The main niche in the Russian energy complex is highlighted, in which its development on the principles of the Industry 4.0 paradigm is possible and most likely.

1. Introduction

The fuel and energy complex plays a crucial role in ensuring the growth rate and resilience of the Russian economy. At the end of 2018, Russia was in first place in the world in terms of oil production, in the second place in gas production, in the fourth place in electricity generation and in the sixth place in coal production. In 2019, the fuel and energy complex provided about 22% of GDP, almost 60% of exports and 40% of budget revenue. According to the latest forecast for the development of the energy sector until 2040, presented by British Petroleum, Russia will remain the leading player in the global energy market in the coming years, and Russian exports will provide more than 5% of the global demand for primary energy by 2040 [1]. Despite the difficult economic environment and the negative attitude towards coal and coal generation amid the struggle to reduce CO2 emissions into the atmosphere, coal will remain an important source of electricity for a long time; Russia will remain its most important producer and exporter, having a real opportunity to expand Russia's presence on the world coal market, strengthen its position and increase its share due to such high-quality competitive advantages as high calorie content, low content of sulfur, nitrogen and ash. Over a 20-year period, Russia's share in international coal trade has increased more than 3.5 times and has amounted to 14% [1].

Recently, the Russian economy and its energy sector have faced a new challenge known as the transition of the world economy to a new development paradigm; this is the industrial revolution, which is called Industry 4.0. It is heavily dependent on manufacturing and high-tech sectors but experts are sure that it is the technologies of Industry 4.0 that will become the basis for the development of the extractive sectors, including the coal industry. It is safe to say that the future lies with those companies that can take advantage of Industry 4.0 [2, 3].

A sufficient amount of work has already appeared analyzing the features of digital technologies in the processing and enrichment industry [4-7]. However, the term Industry 4.0 itself and the set of digital and production technologies describing it vary greatly from article to article [8-10]. This is largely due to the fact that the new paradigm of industrial production (Industry 4.0) is still taking shape. In addition, it is now clear that in different industries it will go in different ways and at different speeds [8, 11]. Most likely, in its final form, it will rely on a kind of mixture of related production and digital technologies that have serious industry and company differences, while having high inter-industry interoperability [12-13].

There are already attempts to predict how Industry 4.0 will look in the mining industries. The authors make predictions about the introduction of unmanned dump trucks and drones in quarries, the introduction of the SX-EW method in copper processing or new filtering technologies. Most of the Industry 4.0 research in the extractive industries focuses on the discussion of new machines and new main processes [14-15]. Already now we can conclude that, unlike previous industrial revolutions, Industry 4.0 does not rely on a single technology. In these conditions, it makes sense to discuss not only the technologies themselves (digital and production), but also the specific points of growth of these technologies and their possible impact on business processes. One approach that can answer these questions is the "value-chain analysis" (VCA) concept.

The purpose of this article is to identify Industry 4.0 growth points in the domestic coal industry and the forecast of possible changes in the business models of energy companies using the concept of value chains.

2. Subject and methods of research

To analyze the current and emerging paradigms of development of the global energy sector, data from world energy statistics posted on the websites of IAE (International Energy Agency) and BP was used. The information on the state of the Russian fuel and energy complex was obtained from the official websites of the Federal State Statistics Service of the Russian Federation and Ministry of Energy of the Russian Federation. To assess the degree of readiness of Russian oil and gas and coal enterprises, the authors of the article analyzed the official documents of companies (reports, development strategies), official websites and public speeches of top managers over the period 2017 to 2019. As a theoretical basis, the work uses the concept of value chains (Value-chain analysis VCA).

3. Industry 4.0 and the upgrade of value chains in the energy sector

The industrial revolution in the energy sector is under the pressure of three main groups of factors: the exhaustion of energy sources that were traditional for the previous stage of development, rapid changes in the energy market and the emergence of new technological development possibilities, including digital and manufacturing technologies (figure 1).

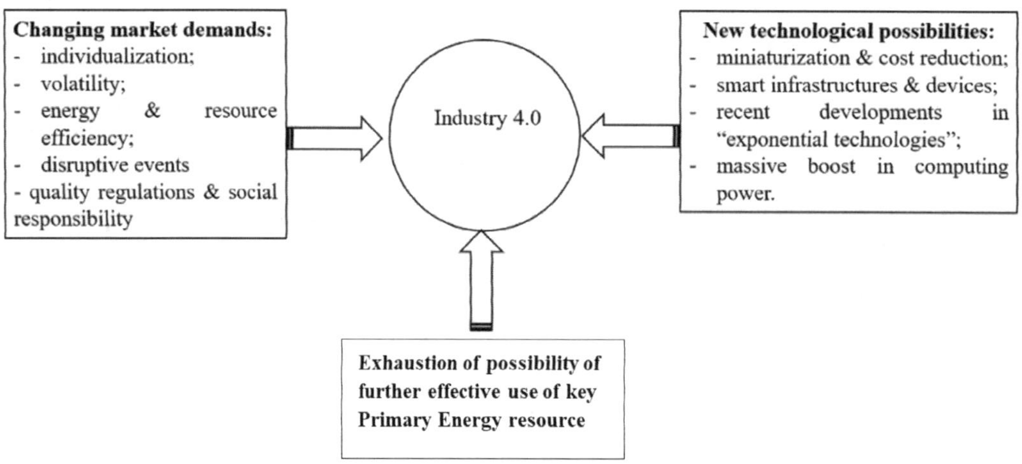

Figure 1. Driving forces of Industry 4.0.

In the history of mankind, the industrial revolution has always relied on a change in the basic source and key technology (see figure 2).

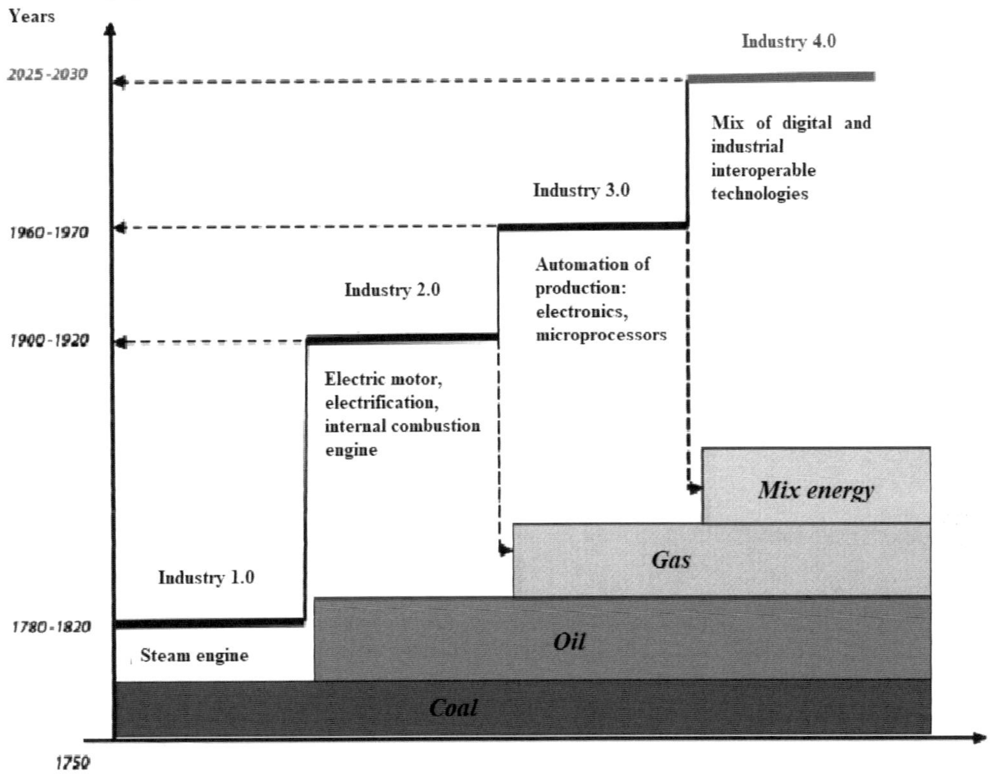

Figure 2. World industrial revolutions and world energy transitions [16].

Industry 1.0 relied on coal and a steam engine; electricity and oil lay at the core of Industry 2.0; Industry 3.0 is associated with active gas development and automation. Industry 4.0, unlike previous revolutions, is not accompanied by a transition to a new energy source, when the fuel balance was achieved by reducing the extraction of one resource and the growth of another; overcoming resource limitations is due to the amplification of reducing energy intensity and the creation of hybrid resources with predetermined properties. In the fuel balance, there is not so much replacement of one resource by another, but diversification that is the equalization of the shares of coal, oil, gas and non-mineral alternative sources of primary energy. The most important requirements for them are the predetermined consumer properties and on time flexible deliveries. This allows concluding even now that in the near future a whole range of hybrid interchangeable resources with predetermined properties will become the leading energy resource.

In addition to changing the basic energy carrier, all previous industrial revolutions were accompanied by a change in key technology that determined the nature and structure of markets, especially supply chains.

In modern foreign and domestic studies, the fourth industrial revolution is mainly considered through the prism of using information and digital technologies in industrial processes. Among them, the most important are Cyber-Physical System (CPS), Internet of Things (IoT) in industry (the usage of digital counterparts, scanners and sensors), Big Data and Business Intelligence, Augmented Reality (AR) technology, Cloud Storage technologies, Autonomous Work, Information Security (secure data, access authentication system and full control of operation networks), Additive Manufacturing, Digital

Modeling and Horizontal and Vertical System Integration (internal and external organizational interaction of the enterprise) [17].

However, it would be wrong to reduce the fourth industrial revolution to digital technology alone. Changes are occurring swiftly; along with digital technologies they are connected, firstly, with breakthrough industry production technologies, which no less than digital ones change the production processes, structure and organization of company management. Secondly, the changes are the result of occurrence of platform solutions, which are based on new products and the way they are promoted and consumed, which means a radical restructuring of the markets and the mechanisms of interaction of the main participants on them. Thirdly, there are changes in the relationship of companies with the external environment, involving the erosion of borders enterprises, industries and sectors. There is also a transformation of part of the external environment of companies into the internal one, which opens up new opportunities for interaction between business, the state and society. The authors of the article define the concept of "Industry 4.0" as a profound change in the economic structure: an interconnected change in technology, organizations, markets and society, which comprehensively affects all sectors and aspects of human life.

The distinguishing feature of Industry 4.0 is that it is difficult to define a particular technology for it. Digitalization, the Internet of Things, advanced robotics, artificial intelligence, machine learning, block chain and the creation of new materials have led to serious technological and structural changes. Automation, sensors, and Internet connection create great potential for increasing efficiency and lowering costs in the energy sector. According to the World Economic Forum, digitalization of the oil and gas industry alone could bring additional revenue of $ 1.6 trillion by 2026 [18]. It is true that digitalization largely determines the nature of the changes in the transition to Industry 4.0. But just as in the situation with energy resources, it is impossible to define one leading technology here. The key here is a mix of production and digital interoperable technologies, while for each individual industry this mixture has distinctive properties.

Finally, new technologies modify markets and impose new requirements on key players, as well as on the rules for their interaction between themselves and companies from related industries. As a result, the nature and structure of the energy market is changing. In the framework of the modern industrial revolution, the producer market has been replaced by the consumer market; the number of consumers has grown rapidly; energy prices have become extremely volatile. The structure of companies is gradually changing. This is especially noticeable when comparing the traditional value chain in the energy sector and the new hybrid cluster formations that arise as a result of upgrading value chains as they transit to Industry 4.0 (figure 3).

In its most general form, the added value chain can be represented as "the whole cycle of the organization, production, and delivery of products from inception to use and recycling which provides a tool for mapping these crucial domains of private and public policy. But, more than that, by focusing on the dynamic shifting of producer rents through the chain, on the processes whereby key actors provide governance to production that occurs on a global basis, value-chain analysis provides important insights into ... challenges confronting both private and public actors " [19].

In most extractive industries, including the coal industry, until recently, AVCs were fairly simple (figure 3a). Most of the economic profit was generated due to the resource component (the quality of coal and the conditions of its bedding), and the management process was reduced to the establishment of highly integrated vertical companies that would create barriers for the competitors to enter the sector and supplement the natural rent with the monopoly one. Now in the energy sector global vertically integrated companies (VICs), whose competitiveness is largely determined by the ability to build upward, supply chains, are being replaced by national and regional companies included in horizontal networks (clusters), whose competitiveness is largely determined by flexibility and ability to cooperate and adapt to flexible market conditions. Large energy companies are increasingly cooperating with companies in high-tech industries. So only during 2017 did BP acquire Beyond Limits, a start-up based on artificial intelligence and cognitive computing that adapts NASA's upstream technologies for deep space exploration to the sector. Chevron is actively developing graphics processors for visualizing seismic data and creating three-dimensional field models. The main goal is to determine the most suitable places for drilling. Shell is developing machine learning algorithms for seismic exploration to automatically detect and classify geological structures in onshore and offshore oil and gas fields. This shows that it's not just a deep transformation of the energy sector, but the barriers between sectors and the borders between suppliers, producers and consumers that are being destroyed. These global changes make it possible to talk about a complete paradigm shift in the development of the energy sector and related industries [20]. They almost completely put an end to the old approach to the extraction and use of energy resources, the essence of which was "the consistent development of new" mineral reserves, with "the main emphasis on the discovery and development of primarily large and giant deposits", and require active inclusion in the process of developing a new conceptual approach. In recent decades, there has been a shift in direct control over global oil and gas resources from "super majors" (the largest transnational non-state oil and gas and coal companies) to national and regional companies [21-22]. All this has affected the value chains in the energy sector, which are gradually turning into integrated value chains (figure 3b).

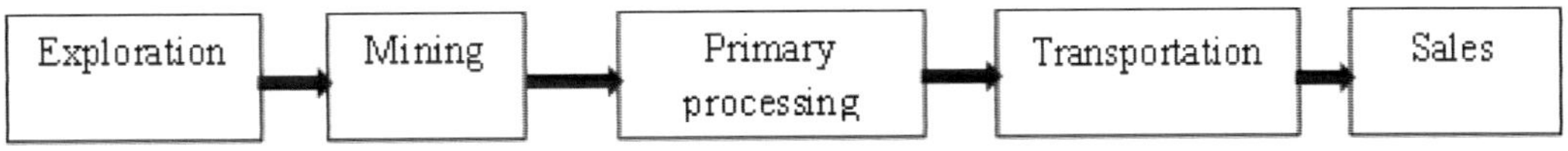

Figure 3. (a) Traditional value chain in the energy sector [18].

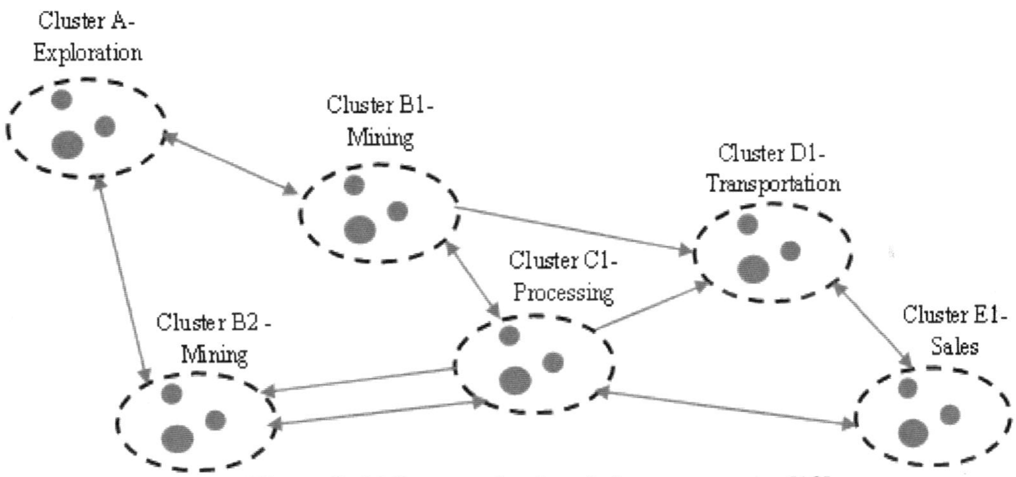

Figure 3. (b) Integrated value chain energy sector [18].

Finally, the business model of companies is changing. The traditional model of energy organization is being replaced by the platform model and the "smart field" model, which in the future can turn into an integral cyber-physical system (table 1).

4. Discussion. Readiness of the Russian energy companies to introduce elements of Industry 4.0 based on digital technology

Until recently, the largest companies in the Russian energy sector have had a vertically integrated structure, which was especially evident in the oil and gas sector. According to the Ministry of Energy of the Russian Federation, 290 organizations carry out oil and gas condensate production in Russia, of which 100 organizations are part of 11 vertically integrated companies (VICs), mainly oil producers, and two are mainly gas and condensate producers. The number of independent mining companies outside the VIC structure is 187. Three companies operate in Russia under production sharing agreements (PSA).

Table 1. Modern concepts of digital business models.

	Platform model	"Smart field" model	Cyber-physical system
	Upgrade (modification) of existing value chains		Creation of new value chains
Key function	Based on market and price management by coordinating the activities of various market participants	Based on full product lifecycle management	Based on the design and management of the entire value chain and the external environment of the company
Key digital business processes	Automatically tracks prices, markets, and consumer requests; Replaces intermediaries - optimizes the process of concluding transactions and conducting payments	Flexible serial automated production – transformer; Lean, automated, supply chain driven by the Internet of things; Integration with research, service, logistics and sales	A single complex of computing resources and physical processes, covering both individual enterprises and complexes of enterprises implementing sequential redistribution in value chains; Formation of a digital eco-system, including business and other interested parties
Key digital technologies	Big data, cloud storage, distributed registries	Industrial Internet of Things, robots, including drones, digital doubles, smart materials, 3D printing	
Removed restrictions	Expanding the boundaries of demand constraints	Expanding resource limits	Formation of a new development paradigm based on the principles of Industry 4.0

In the structure of oil and condensate production, VICs account for 80.3%; independent companies produce 16.4% of hydrocarbons, and PSA companies accounts for 3.4% of the production. The share of state-owned companies (Rosneft PJSC and Gazprom Neft PJSC, excluding unconsolidated assets) amounts to 42%, which is by 0.3% more than the previous year [22].

Over the past years, oil and gas companies have experienced severe resource constraints. Mega large fields with readily available resources are gradually exhausting themselves; over the past 10 years, independent oil companies operating with small and medium fields and the so-called hard-to-recover reserves have shown stable positive growth dynamics in oil production in Russia. They formally do not enter the structure of VICs. The average growth rate of oil production by independent companies over the past 10 years has been 8.2%, and the growth in production over the previous year was 2.1 million tons.

Changing the structure of the mineral resource base, the growth of hard-to-recover oil and gas reserves necessitate the development of new technological and organizational solutions. The search, exploration and production processes are significantly more expensive in connection with the development of deposits in remote and infrastructure-undeveloped places, the shelf of the Arctic seas, which requires the adaptation of known technologies to unconventional hydrocarbon production conditions. In the 1990's, there was a demand in the industry for innovative process control tools to ensure the economic efficiency of oil and gas production in these difficult conditions. Therefore, the digitalization of the oil and gas industry is mainly related to the development of the "smart field" business model related to the development of forms of automation and management of the process of extraction, transportation and hydrocarbons processing [22].

One of the first digital projects in the oil and gas industry was the "smart well" project, implemented in 1997 at the Snorre Field in the North Sea by the Norwegian company Saga Petroleum. Later, the term was expanded to "smart field", which allowed it to be applied to larger objects. The so-called "smart" technologies have been developed and disseminated throughout the AVC in the oil and gas industry. The effectiveness of these digital technologies is achieved by increasing the level of control and automation of the production process, as well as minimizing labor. However, the latter is controversial because most likely there is a change in the structure of employees, and often towards downgrading [22].

Currently, the terms "intellectual field" (i-field), "field of the future" and "digital oil field of the future" (DOFF) are widely used. The purpose of creating "intelligent" deposits is the ability to ensure transparency and accessibility of the necessary production, geological, economic and other parameters and indicators of the enterprise [21].

In addition, the development of digital systems has allowed the formation in the oil and gas industry of a new element of the AVC – "administrative" or "managerial". Thus, following high technology, the organizational structure of companies is being transformed, and the centralization process is intensifying. Subsidiary production units of large oil and gas companies are limited or completely deprived of their freedom of decision-making, since data processing and the formation of a development strategy take place in the head unit (holding) [22].

But, despite the existing positive aspects of the digitalization of the oil and gas industry, its level of Industry 4.0 formation lags far behind the world [23].

Somewhat different, the processes of Industry 4.0 are developing in the Russian coal industry. This is due to the fact that the domestic coal industry is much less monopolized than the oil and gas sector. This is because of the fact that they were privatized, and the enterprises incorporated in them became part of the energy, metallurgical, chemical and construction holdings as one of the lower branches. Due to this, such large monopolists as Gazprom and Rosneft did not survive in the coal industry; the AVCs turned out to be shorter and more branched. However, in the coal industry, most of the companies are still united in vertically integrated holdings, working mainly for export [24].

Another important factor, determining the features of the formation of Industry 4.0 and the creation AVCs in the industry, is more stringent demand constraints in the world market compared to the oil and gas sector. The fall in coal prices in 2011 and the decarbonization policy, pursued in developed countries, led to the fact that digital platform solutions in the field of logistics and sales began to actively develop in the coal industry. This made it possible to increase the efficiency of coal mining, reduce operating costs, and optimize other indicators of industry enterprises. Separate elements of Industry 4.0 and digitalization are also present in coal mining and processing technologies. There has appeared robotic or remotely controlled equipment, remote centralized control rooms, high-precision positioning of drilling rigs or excavators, portable devices for monitoring personnel health status, three-dimensional design and management enterprise models, predictive methods for analyzing geological data and everything related to the production process. However, they are mainly of a demonstration or experimental nature. The consequence of the energy transition and digitalization of the industry has been the growing role of medium and small independent coal mining companies focused on local markets [25].

The transition to digital technologies has created the conditions for cooperation between these companies within the framework of cooperative AVCs with the involvement of scientific and educational organizations and the state. The importance of intersectoral cooperation in the energy sector has grown. Coal producers and consumers are forced to adapt to the extreme volatility of markets and diversify their business, including the incorporation of small coal companies operating in local markets in their orbit of influence [16].

With the digitalization of logistics and the transport sector, the AVCs of large metallurgical and especially energy holdings have begun to change. Management and profit centers in these AVCs have begun to move to engineering, logistics and transport companies. AVCs themselves have become more flexible and decentralized. Even in developing countries, where with the support of the state vertically integrated coal companies with centralized AVCs maintained for especially long, new forms of competition have started to develop gradually; this sector has started to more and more join the "network economy". At the same time, unlike the oil and gas sector, which is developing towards the creation of a "smart field" business model, the Russian coal mining companies are mainly going along the path of creating platform solutions in the field of logistics. In this regard, the experience of SUEK, the largest Russian and world coal-mining holding company, is very indicative [26].

However, in general, the digitalization level of the Russian coal industry is still far behind the digitalization level of the coal industries of foreign countries such as the USA, the UK, Japan, the Czech Republic and Germany. The processes of digitalization and transformation of the industry are point-like in nature, have little impact on the areas of logistics and transport and have little effect on downstream costs.

5. Conclusion

The intensification of the global energy complex necessitates the search for a new paradigm for the development of the Russian fuel and energy complex with a focus on new approaches to the development of the raw material base and energy transition to environmentally friendly, mixed and renewable energy sources, sustainable development through the use of integrated geotechnologies and the formation of new added value chains.

The Industry 4.0 concept of industrial development opens up new opportunities for enterprises of the fuel and energy complex of Russia to significantly increase labor productivity, ensure safety, and rational subsoil use. Thanks to digitalization technologies, the "Internet of Things", artificial intelligence and elements of robotization of the main and auxiliary technological processes, significant potential is being formed to reduce costs and increase the competitiveness of the Russian fuel and energy complex in the global energy market.

Acknowledgements

This article was prepared with the support of Russian Science Foundation (No. 17-78-20218, the project "Spatial specialization and holistic development of resource-type regions").

References

[1] BP Outlook 2019 www.bp.com/en/global/corporate/energy-economics/energy-outlook.html
[2] Klaus Schwab 2016 *World Economic Forum* (Geneva, Switzerland) p 184
[3] Paul H 2016 Industrie 4.0. *Annäherung an ein Konzept* (Dortmund: Gelsenkirchen: IAT)
[4] Jämsä-Jounela S-L 2007 *Future Trends in Process Automation. Annual Reviews in Control* 31(2) 211–220
[5] Groenveld D and Rozou M 2016 *Advanced Process Control for Grinding Circuits*
[6] Ofstie D and Kargutkar B *Proc. 54th Annual Conference of Metallurgists hosting America's Conference on Aluminium Alloys* (Canada: Canadian Institute of Mining Metallurgy and Petroleum)
[7] Van Duin S, Meers L and Gibson G 2013 *Proc. of 30th International Symposium on Automation and Robotics in construction and Mining* (ISARC 2013) (Montreal, Quebec, Canada)
[8] Bassi L 2017 *IEEE 3rd International Forum on Research and Technologies for Society and Industry – Innovation to Shape the Future for Society and Industry (RTSI)* (Piscataway, NJ: IEEE) pp 1–6
[9] Chiarello F et al 2018 *Computers in Industry* **100** 244–257
[10] *Centre for the Fourth Industrial Revolution Network for Global Technology Governance. Web Economic Forum* 2019 p 24
[11] Stanway G et al 2015 *The Australasian Institute of Mining and Metallurgy* ed Iron Ore Conference (Perth) pp 631–645
[12] Goosen E V, Nikitenko S M et al 2019 *Eurasian Mining* **2** 36-40
[13] Nikitenko S M, Goosen E V et al 2017 *Fundamental Research* 2017 no 10 part 2 375–380
[14] Lumley L 2012 *The Australasian Institute of Mining and Metallurgy ed Project Evaluation Conference* pp 203–224
[15] Stanway G et al 2015 *The Australasian Institute of Mining and Metallurgy ed Iron Ore Conference* (Perth) pp 631–645

[16] Plakitkin Yu A and Plakitkina L S 2017 *Ugol* 2017 **10(1099)**
[17] Camarinha-Matos L M, Fornasiero R and Afsarmanesh H 2017 *Collaboration in a Data-Rich World. PRO-VE 2017. IFIP Advances in Information and Communication Technology* **506** 3-17
[18] *Capturing Value from Disruption Technology and Innovation in an Era of Energy Transformation. PwC global power & utilities. PwC* 2016 p 36
[19] Kaplinsky R 2004 *Journal Problems of Economic Transition* vol 474 iss 2 74-115
[20] *The Industry's Future: The Destruction of Barriers. PwC Series of Development Prospects PwC* 2018 p 16
[21] Kontorovich A E 2016 *Science First-hand* **1(67)** 6-17
[22] Kontorovich A E, Eder L V and Filimonova I V 2017 *IOP Conference Series: EES* **84** 012010
[23] *Competing in the Digital Age. Strategic Challenges for the Russian Federation 2018* http://documents.worldbank.org
[24] Nikitenko S M and Goosen E V 2017 *ECO* vol 9 (519) 104–124
[25] Pestun V 2019 *Jet Info* **5-6**
[26] *Capturing Value from Disruption Technology and Innovation in an Era of Energy Transformation. PwC global power & utilities. PwC* 2016 p 36

AUTHOR INDEX